Intelligent Autonomous Robotics
A Robot Soccer Case Study

Synthesis Lectures on Artificial Intelligence and Machine Learning

Editors
Ronald J. Brachman, *Yahoo Research*

Tom Dietterich, *Oregon State University*

Intelligent Autonomous Robotics
Peter Stone
2007

Intelligent Autonomous Robotics

Peter Stone, University of Texas at Austin

ISBN: 978-3-031-00416-2   paperback
ISBN: 978-3-031-00416-2   paperback

ISBN: 978-3-031-01544-1   ebook
ISBN: 978-3-031-01544-1   ebook

DOI: 10.1007/978-3-031-01544-1

A Publication in the Springer series
SYNTHESIS LECTURES ON ARTIFICIAL INTELLIGENCE AND MACHINE LEARNING #1
Lecture #1
Series Editors : Ronald Brachman, Yahoo! Research and Thomas G. Dietterich, Oregon State University

First Edition
10 9 8 7 6 5 4 3 2 1

Intelligent Autonomous Robotics
A Robot Soccer Case Study

Peter Stone
University of Texas at Austin

SYNTHESIS LECTURES ON ARTIFICIAL INTELLIGENCE AND MACHINE LEARNING #1

ABSTRACT

Robotics technology has recently advanced to the point of being widely accessible for relatively low-budget research, as well as for graduate, undergraduate, and even secondary and primary school education. This lecture provides an example of how to productively use a cutting-edge advanced robotics platform for education and research by providing a detailed case study with the Sony AIBO robot, a vision-based legged robot. The case study used for this lecture is the UT Austin Villa RoboCup Four-Legged Team. This lecture describes both the development process and the technical details of its end result. The main contributions of this lecture are (i) a roadmap for new classes and research groups interested in intelligent autonomous robotics who are starting from scratch with a new robot, and (ii) documentation of the algorithms behind our own approach on the AIBOs with the goal of making them accessible for use on other vision-based and/or legged robot platforms.

KEYWORDS

Autonomous robots, Legged robots, Multi-Robot Systems, Educational robotics, Robot soccer, RoboCup

Contents

CHAPTER 1

Introduction

Robotics technology has recently advanced to the point of being widely accessible for relatively low-budget research, as well as for graduate, undergraduate, and even secondary and primary school education. However, for most interesting robot platforms, there remains a substantial learning curve or "ramp-up cost" to learning enough about the robot to be able to use it effectively. This learning curve cannot be easily eliminated with published curricula or how-to guides, both because the robots tend to be fairly complex and idiosyncratic, and, more importantly, because robot technology is advancing rapidly, often making previous years' models obsolete as quickly as competent educational guides can be created.

This lecture provides an example of how to productively use a cutting-edge advanced robotics platform for education and research by providing a detailed case study with the Sony AIBO robot. Because the AIBO is (i) a legged robot with primarily (ii) vision-based sensing, some of the material will be particularly appropriate for robots with similar properties, both of which are becoming increasingly prevalent. However, more generally, the lecture will focus on the steps required to start with a new robot "out of the box" and quickly using it for education and research.

The case study used for this lecture is the UT Austin Villa RoboCup Four-Legged Team. In 2003, UT Austin Villa was a new entry in the ongoing series of RoboCup legged league competitions. The team development began in mid-January of 2003, at which time none of the team members had any familiarity with the AIBOs. Without using any RoboCup-related code from other teams, we entered a team in the American Open competition at the end of April, and met with some success at the annual RoboCup competition that took place in Padova, Italy, at the beginning of July. By 2004, the team became one of the top teams internationally, and started generating a series of research articles in competitive conferences and journals.

RoboCup, or the Robot Soccer World Cup, is an international research initiative designed to advance the fields of robotics and artificial intelligence by using the game of soccer as a substrate challenge domain [3, 6, 39, 41, 52, 54, 57, 77, 90]. The long-term goal of RoboCup is, by the year 2050, to build a full team of 11 humanoid robot soccer players that can beat

FIGURE 1.1: An image of the AIBO and the field. The robot has a field-of-view of 56.9° (hor) and 45.2° (ver), by which it can use the two goals and four visually distinct beacons at the field corners for the purposes of localization.

the best human soccer team on a real soccer field [42]. RoboCup is organized into several different leagues, including a computer simulation league and two leagues that use wheeled robots. The case study presented in this lecture concerns the development of a new team for the Sony four-legged league[1] in which all competitors use identical Sony AIBO robots and the Open-R software development kit.[2] Here, teams of four AIBOs, equipped with vision-based sensors, play soccer on a color-coded field. Figure 1.1 shows one of the robots along with an overhead view of the playing field. As seen in the diagram, there are two goals, one at each end of the field and there is a set of visually distinct beacons (markers) situated at fixed locations around the field. These objects serve as the robot's primary visual landmarks for localization.

The Sony AIBO robot used by all the teams is roughly 280 mm tall (head to toe) and 320 mm long (nose to tail). It has 20 degrees of freedom: 3 in its head, 3 in each leg, and 5 more in its mouth, ears, and tail. It is equipped with a CMOS color camera at the tip of its nose with a horizontal field-of-view of 56.9° and a vertical field-of-view of 45.2°. Images are captured at 30 frames per second in the YCbCr image format. The robot also has a wireless

---

[1]http://www.tzi.de/4legged/.

[2]http://openr.aibo.com/.

LAN card that allows for communication with other robots or an off-board computer. All processing is performed on-board the robot, using a 576 MHz processor.[3] Since all teams use identical robots, the four-legged league amounts to essentially a software competition.

This lecture details both the development process and the technical details of its end result, a new RoboCup team, called UT Austin Villa,[4] from the Department of Computer Sciences at the University of Texas at Austin. The main contributions are

1. A roadmap for new classes and research groups interested in intelligent autonomous robotics who are starting from scratch with a new robot; and

2. Documentation of the algorithms behind our own approach on the AIBOs with the goal of making them accessible for use on other vision-based and/or legged robot platforms.

As a case study, this lecture contains significant material that is motivated by the specific robot soccer task. However, the main general feature of the class and research program described is that there was a concrete task-oriented goal with a deadline. Potential tasks other than soccer include autonomous surveillance [1, 56], autonomous driving [50], search and rescue [51], and anything else that requires most of the same subtask capabilities as robot soccer as described in Chapter 2.

Though development on the AIBOs has continued in our group for several years after the initial ramp-up, this lecture focuses extensively on the first year's work as an example of starting up education and research on a new robot from scratch. Some of the later years' developments are also documented as useful and appropriate.

In the four-legged league, as in all RoboCup leagues, the rules are changed over time to make the task incrementally more difficult. For example, in the first year of competition documented in this lecture (2003), the field was 2.9 m × 4.4 m and there were walls surrounding the field. By 2005, the field had been enlarged to 3.6 m × 5.4 m and the walls were removed. As such, some of the images and anecdotes in this lecture reflect slightly different scenarios. Nonetheless, the basic flow of games has remained unchanged and can be summarized as follows.

- Teams consist of four robots each;
- Games consist of two 10-minute halves with teams switching sides and uniform colors at half-time;

---

[3]These specifications describe the most recent ERS-7 model. Some of the details described in this lecture pertain to the early ERS-210A that was slightly smaller, had slightly less image resolution, and a somewhat slower processor. Nonetheless, from a high level, most of the features of these two models are similar.

[4]http://www.cs.utexas.edu/~AustinVilla.

- Once the play has started, the robots must operate fully autonomously, with no human input or offboard computation;

- The robots may communicate via a wireless LAN;

- If the ball goes out of bounds, a human referee returns it to the field;

- No defenders (other than the goalie) are allowed within the goalie box near the goal;

- Robots may not run into other robots repeatedly;

- Robots may not grasp for longer than 3 s;

- Robots that violate the rules are penalized by being removed from the field for 30 s, after which they are replaced near the middle of the field;

- At the end of the game, the team that has scored the most goals, wins.

Since some of these rules rely on human interpretation, there have been occasional arguments about whether a robot should be penalized (sometimes hinging around what the robot "intended" (!) to do). But, for the most part, they have been effectively enforced and adhered to in a sportsmanlike way. Full rules for each year are available online at the four-legged-league page cited above.

The following chapter outlines the structure of the graduate research seminar that was offered as a class during the Spring semester of 2003 and that jump-started our project. At the end of that chapter, I outline the structure of the remainder of the lecture.

CHAPTER 2

The Class

The UT Austin Villa legged robot team began as a focused class effort during the Spring semester of 2003 at the University of Texas at Austin. Nineteen graduate students, and one undergraduate student, were enrolled in the course CS395T: *Multi-Robot Systems: Robotic Soccer with Legged Robots*.[1]

At the beginning of the class, neither the students nor the professor (myself) had any detailed knowledge of the Sony AIBO robot. Students in the class studied past approaches to four-legged robot soccer, both as described in the literature and as reflected in publicly available source code. However, we developed the entire code base *from scratch* with the goals of learning about all aspects of robot control and of introducing a completely new code base to the community.

Class sessions were devoted to students educating each other about their findings and progress, as well as coordinating the integration of everybody's code. Just nine weeks after their initial introduction to the robots, the students already had preliminary working solutions to vision, localization, fast walking, kicking, and communication.

The concrete goal of the course was to have a completely new working solution by the end of April so that we could participate in the RoboCup American Open competition, which happened to fall during the last week of the class. After that point, a subset of the students continued working towards RoboCup 2003 in Padova.

The class was organized into three phases. Initially, the students created simple behaviors with the sole aim of becoming familiar with Open-R.

Then, about two weeks into the class, we shifted to phase two by identifying key subtasks that were important for creating a complete team. Those subtasks were

- Vision;
- Movement;
- Fall Detection;
- Kicking;

[1] http://www.cs.utexas.edu/~pstone/Courses/395Tspring03.

- Localization;

- Communication;

- General Architecture; and

- Coordination.

During this phase, students chose one or more of these subtasks and worked in subgroups on generating initial solutions to these tasks in isolation.

By about the middle of March, we were ready to switch to phase three, during which we emphasized "closing the loop," or creating a single unified code-base that was capable of playing a full game of soccer. We completed this integration process in time to enter a team in the RoboCup American Open competition at the end of April.

The remainder of the lecture is organized as follows. Chapter 3 documents some of the initial behaviors that were generated during phase one of the class. Next, the output of some of the subgroups that were formed in phase two of the class, is documented in Chapters 4–8. Next, the tasks that occupied phase three of the class are documented, namely those that allowed us to put together the above modules into a cohesive code base (Chapters 9–13). Chapters 14 and 15 introduce our simulator and debugging and development tools, and Chapter 16 concludes. In all chapters, emphasis is placed on the general lessons learned, with some of the more AIBO-specific details left for the appendices.

CHAPTER 3

Initial Behaviors

The first task for the students in the class was to learn enough about the AIBO, to be able to compile and run any simple program on the AIBO.

The open source release of Open-R came with several sample programs that could be compiled and loaded onto the AIBO right away. These programs could do simple tasks such as

L-Master-R-Slave: Cause the right legs to mirror manual movements of the left legs.
Ball-Tracking-Head: Cause the head to turn such that the pink ball is always in the center of the visual image (if possible).
PID control: Move a joint to a position specified by the user by typing in a telnet window.

The students were to pick any program and modify it, or combine two programs in any way. The main objective was to make sure that everyone was familiar with the process for compiling and running programs on the AIBOs. Some of the resulting programs included the following.

- Variations on L-Master-R-Slave in which different joints were used to control each other. For example, one student used the tail as the master to control all four legs, which resulted in a swimming-type motion. Doing so required scaling the range of the tail joints to those of the leg joints appropriately.

- Variations on Ball-Tracking-Head in which a different color was tracked. Two students teamed up to cause the robot to play different sounds when it found or lost the ball.

- Variations on PIDcontrol such that more than one joint could be controlled by the same input string.

After becoming familiar with the compiling and uploading process, the next task for the students was to become more familiar with the AIBO's operating system and the Open-R interface. To that end, they were required to create a program that added at least one new subject–observer connection to the code.[1] The students were encouraged to create a new Open-R object from scratch. Pattern-matching from the sample code was encouraged, but creating an object as different as possible from the sample code was preferred.

Some of the responses to this assignment included the following.

- The ability to turn on and off LEDs by pressing one of the robots' sensors.
- A primitive walking program that walks forward when it sees the ball.
- A program that alternates blinking the LEDs and flapping the ears.

After this assignment, which was due after just the second week of the class, the students were familiar enough with the robots and the coding environment to move on to their more directed tasks with the aim of creating useful functionality.

---

[1]A subject–observer connection is a pipe by which different Open-R objects can communicate and be made interdependent. For example, one Open-R object could send a message to a second object whenever the back sensor is pressed, causing the second object to, for example, suspend its current task or change to a new mode of operation.

CHAPTER 4

Vision

The ability of a robot to sense its environment is a prerequisite for any decision making. Robots have traditionally used mainly range sensors such as sonars and laser range finders. However, camera and processing technology has recently advanced to the point where modern robots are increasingly equipped with vision-based sensors. Indeed on the AIBO, the camera is the main source of sensory information, and as such, we placed a strong emphasis on the vision component of our team.

Since computer vision is a current area of active research, there is not yet any perfect solution. As such, our vision module has undergone continual development over the course of this multi-year project. This lecture focusses on the progress made during our first year as an example of what can be done relatively quickly. During that time, the vision reached a sufficient level to support all of the localization and behavior achievements described in the rest of this lecture. Our progress since the first year is detailed in our 2004 and 2005 team technical reports [79, 80], as well as a series of research papers [71–73, 76].

Our vision module processes the images taken by the CMOS camera located on the AIBO. The module identifies colors in order to recognize objects, which are then used to localize the robot and to plan its operation.

Our visual processing is done using the established procedure of color segmentation followed by object recognition. Color segmentation is the process of classifying each pixel in an input image as belonging to one of a number of predefined color classes based on the knowledge of the ground truth on a few training images. Though the fundamental methods employed in this module have been applied previously (both in RoboCup and in other domains), it has been built from scratch like all the other modules in our team. Hence, the implementation details provided are our own solutions to the problems we faced along the way. We have drawn some of the ideas from the previous technical reports of CMU [89] and UNSW [9]. This module can be broadly divided into two stages: (i) low-level vision, where the color segmentation and region building operations are performed and (ii) high-level vision, wherein object recognition is accomplished and the position and bearing of the various objects in the visual field are determined.

The problem dealt within this chapter differs from more traditional computer vision research in two important ways.

- First, most state-of-the-art approaches to challenging computer vision problems, such as segmentation [14, 55, 69, 85], blob clustering [28, 36], object recognition [5, 68, 88], and illumination invariance [24, 25, 65] require a substantial amount of computational and/or memory resources, taking advantage of multiple processors and/or processing each image for seconds or even minutes. However, robotic systems, such as ours typically have strict constraints on the resources available, but still demand real-time processing. Indeed, in order to take advantage of all the images available to it, we must enable the AIBO to process each one in roughly 33 ms on its single 576 MHz processor.

- Second, most vision algorithms assume a stationary or slowly (infrequently) moving camera [22, 88]. However, mobile robot platforms such as ours are characterized by rapid movements resulting in jerky nonlinear motion of the camera. These are the more pronounced in legged robots as opposed to wheeled robots.

The remainder of this chapter presents detailed descriptions of the subprocesses of our overall vision system. But first, for the sake of completeness, a brief overview of the AIBO robot's CMOS color camera is presented. The reader who is not interested in details of the AIBO robot can safely skip to Section 4.2.

4.1   CAMERA SETTINGS

The AIBO comes equipped with a CMOS color camera that operates at a frame rate of 30 fps. Some of its other preset features are as follows.

- Horizontal viewing angle: 57.6°.
- Vertical viewing angle: 47.8°.
- Lens aperture: 2.0.
- Focal length: 2.18 mm.

We have partial control over three parameters, each of which has three options from which to choose:

- *White balance*: We are provided with settings corresponding to three different light temperatures.

1. *Indoor-mode*: 2800 K.
2. *FL-mode*: 4300 K.
3. *Outdoor-mode*: 7000 K.

This setting, as the name suggests, is basically a color correction system to accommodate varying lighting conditions. The idea is that the camera needs to identify the 'white point' (such that white objects appear white) so that the other colors are mapped properly. We found that this setting does help in increasing the separation between colors and hence helps in better object recognition. The optimum setting depends on the 'light temperature' registered on the field (this in turn depends on the type of light used, i.e., incandescent, fluorescent, etc.). For example, in our lab setting, we noticed a better separation between orange and yellow with the *Indoor* setting than with the other settings. This helped us in distinguishing the orange ball from the other yellow objects on the field such as the goal and sections of the beacons.

- *Shutter Speed*:

1. Slow: 1/50 s.
2. Mid: 1/100 s.
3. Fast: 1/200 s.

This setting denotes the time for which the shutter of the camera allows light to enter the camera. The higher settings (larger denominators) are better when we want to *freeze* the action in an image. We noticed that both the 'Mid' and the 'Fast' settings did reasonably well though the 'Fast' setting seemed the best, especially considering that we want to capture the motion of the ball. Here, the lower settings would result in blurred images.

- *Gain:*

1. Low: 0 dB.
2. Mid: 0 dB .
3. High: 6 dB.

This parameter sets the camera gain. In this case, we did not notice any major difference in performance among the three settings provided.

4.2   COLOR SEGMENTATION

The image captured by the robot's camera, in the YCbCr format, is a set of numbers, ranging from 0 to 255 along each dimension, representing luminance (Y) and chrominance (Cb, Cr). To enable the robot to extract useful information from these images, the numbers have to be suitably mapped into an appropriate color space. We retain the YCbCr format and "train" the robot, using a Nearest Neighbor (NNr) scheme [9, 15], to recognize and distinguish between

10 different colors, numbered as follows:

- 0 = pink;
- 1 = yellow;
- 2 = blue;
- 3 = orange;
- 4 = marker green;
- 5 = red;
- 6 = dark (robot) blue;
- 7 = white;
- 8 = field green; and
- 9 = black.

The motivation behind using the NNr approach is that the colors under consideration overlap in the YCbCr space (some, such as orange and yellow, do so by a significant amount). Unlike other common methods that try to divide the color space into cuboidal regions (or a collection of planes), the NNr scheme allows us to learn a color table where the individual blobs are defined more precisely.

The original color space has three dimensions, corresponding to the Y, Cb, and Cr channels of the input image. To build the color table (used for classification of the subsequent images on the robot), we maintain three different types of color cubes in the training phase: one Intermediate (IM) color cube corresponding to each color, a Nearest Neighbor cube, and a Master (M) cube (the names will make more sense after the description given below). To reduce storage requirements, we operate at half the resolution, i.e., all the cubes have their numerical values scaled to range from 0 to 127 along each dimension. The cells of the IM cubes are all initialized to zero, while those of the NNr cube and the M cube are initialized to nine (the color black, also representing background).

Color segmentation begins by first training on a set of images using UT Assist, our Java-based interface/debugging tool (for more details, see Chapter 15). A robot is placed at a few points on the field. Images are captured and then transmitted over the wireless network to a remote computer running the Java-based server application. The objects of interest (goals, beacons, robots, ball, etc.) in the images are manually "labeled" as belonging to one of the color classes previously defined, using the Image Segmenter (see Chapter 15 for some pictures showing the labeling process). For each pixel of the image that we label, the cell determined by the corresponding YCbCr values (after transforming to half-resolution), in the corresponding IM cube, is incremented by 3 and all cells a certain Manhattan distance away (within 2 units)

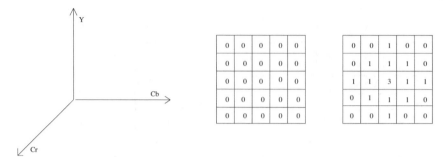

FIGURE 4.1: An example of the development of the color table, specifically the IM cube. Part (a) shows the general coordinate frame for the color cubes. Part (b) shows a planar subsection of one of the IM cubes before labeling. Part (c) depicts the same subsection after the labeling of a pixel that maps to the cell at the center of the subsection. Here only one plane is shown—the same operation occurs across all planes passing through the cell under consideration such that all cells a certain Manhattan distance away from this cell are incremented by 1.

from this cell are incremented by 1. For example, if we label a pixel on the ball orange in the image and this pixel corresponds to a cell $(115, 35, 60)$ based on the intensity values of that pixel in the image, then in the orange IM cube this cell is incremented by 3 while the cells such as $(115, 36, 61)$ and $(114, 34, 60)$ (among others) which are within a Manhattan distance of 2 units from this cell, in the orange IM cube alone, are incremented by 1. For another example, see Fig. 4.1.

The training process is performed incrementally, so at any stage we can generate a single cube (the NNr cube is used for this purpose) that can be used for segmenting the subsequent images. This helps us to see how "well-trained" the system is for each of the colors and serves as a feedback mechanism that lets us decide which colors need to be trained further. To generate the NNr cube, we traverse each cell in the NNr cube and compare the values in the corresponding cell in each of the IM cubes and assign to this cell the index of the IM cube that has the maximum value in this cell, i.e., $\forall (p, q, r) \in [0, 127]$,

$$\text{NNrCube}(y_p, cb_q, cr_r) = \underset{i \in [0,9]}{\text{argmax}}\, \text{IM}_i(y_p, cb_q, cr_r). \tag{4.1}$$

When we use this color cube to segment subsequent images, we use the NNr scheme. For each pixel in the test image, the YCbCr values (transformed to half-resolution) are used to index into this NNr cube. Then we compute the weighted average of the value of this cell and those cells that are a certain Manhattan distance (we use 2–3 units) around it to arrive at a value that is set as the "numerical color" (i.e. the color class) of this pixel in the test image. The weights are proportional to the Manhattan distance from the central cell, i.e., the greater

9	3	1	1	9
3	1	3	3	3
3	3	1	3	3
1	1	3	3	3
9	1	3	3	9

(a)

9	3	1	1	9
3	1	3	3	3
3	3	3	3	3
1	1	3	3	3
9	1	3	3	9

(b)

FIGURE 4.2: An example of the weighted average applied to the NNr cube (a two-dimensional representative example). Part (a) shows the values along a plane of the NNr cube before the NNr scheme is applied to the central cell. Part (b) shows the same plane after the NNr update for its central cell. We are considering cells within a Manhattan distance of 2 units along the plane. For this central cell, color label 1 gets a vote of $3 + 1 + 1 + 1 = 6$ while label 3 gets a vote of $2 + 2 + 2 + 2 + 1 + 1 + 1 + 1 + 1 = 13$ which makes the central cell's label = 3. This is the value that is set as the classification result. This is also the value that is stored in the cell in the M cube that corresponds to the central cell.

this distance, the smaller the significance attached to the value in the corresponding cell (see Fig. 4.2).

We do the training over several images (around 20–30) by placing the robot at suitable points on the field. The idea here is to train on images that capture the beacons, goals, ball, and the robots from different distances (and also different angles for the ball) to account for the variations in lighting along different points on the field. This is especially important for the orange ball, whose color could vary from orange to yellow to brownish-red depending on the amount of lighting available at that point. We also train with several different balls to account for the fact that there is a marked variation in color among different balls. At the end of the training process, we have all the IM cubes with the corresponding cells suitably incremented. The NNr operation is computationally intensive to perform on the robot's processor. To overcome this, we precompute the result of performing this operation (the Master Cube is used for this) from the corresponding cells in the NNr color Cube, i.e., we traverse each cell of the M Cube and compute the "Nearest Neighbor" value from the corresponding cells in the NNr cube. In other words, $\forall (p, q, r) \in [0, 127]$ with a predefined Manhattan distance $\text{ManDist} \in [3, 7]$,

$$\text{MCube}\,(y_p, cb_q, cr_r) = \underset{i \in [0,9]}{\arg\max}\,\text{Score}(i) \qquad (4.2)$$

where $\forall(k_1, k_2, k_3) \in [0, 127]$,

$$\text{Score}(i) = \left(\sum_{k_1, k_2, k_3} \left(\text{ManDist} - (\mid k_1 - p \mid + \mid k_2 - q \mid + \mid k_3 - r \mid) \right) \right) \mid$$

$$(\mid k_1 - p \mid + \mid k_2 - q \mid + \mid k_3 - r \mid) < \text{ManDist}$$
$$\wedge \quad \text{NNrCube}(y_{k_1}, cb_{k_2}, cr_{k_3}) = i. \qquad (4.3)$$

This cube is loaded onto the robot's memory stick. The pixel-level segmentation process is reduced to that of a table lookup and takes ≈ 0.120 ms per image. For an example of the color segmentation process and the Master Cube generated at the end of it, see Fig. 16.1.

One important point about our initial color segmentation scheme is that we do not make an effort to normalize the cubes based on the number of pixels (of each color) that we train on. So, if we labeled a number of yellow pixels and a relatively smaller number of orange pixels, then we would be biased towards yellow in the NNr cube. This is not a problem if we are careful during the training process and label regions such that all colors get (roughly) equal representation.

Previous research in the field of segmentation has produced several good algorithms, for example, mean-shift [14] and gradient-descent based cost-function minimization [85]. But these involve more computation than is feasible to perform on the robots. A variety of previous approaches have been implemented on the AIBOs in the RoboCup domain, including the use of decision trees [8] and the creation of axis-parallel rectangles in the color space [12]. Our approach is motivated by the desire to create fully general mappings for each YCbCr value [21].

4.3   REGION BUILDING AND MERGING

The next step in vision processing is to find contiguous *blobs* of constant colors, i.e., we need to *cluster* pixels of the same color into meaningful groups. Though past research in this area has resulted in some good methods [28, 36], doing this efficiently and accurately is challenging since the reasoning is still at the pixel level. Computationally, this process is the most expensive component of the vision module that the robot executes.

The Master cube is loaded onto the robot's memory stick and this is used to segment the images that the robot's camera captures (in real-time). The next step in low-level processing involves the formation of rectangular bounding boxes around connected regions of the same color. This in turn consists of run-length encoding (RLE) and region merging [29, 58], which are standard image processing approaches used previously in the RoboCup domain [89].

As each image is segmented (during the first scan of the image), left to right and top to bottom, it is encoded in the form of run-lengths along each horizontal scan line, i.e., along each line, we store the (x, y) position (the root node) where a sequence of a particular color starts and the number of pixels until a sequence of another color begins. The data corresponding to each run-length are stored in a separate data structure (called RunRegion) and the run-lengths are all stored as a linked list. Each RunRegion data structure also stores the corresponding color. Further, there is a bounding box corresponding to each RunRegion/run-length, which during the first pass is just the run-length itself, but has additional properties such as the number of run-lengths enclosed, the number of actual pixels enclosed, the upper-left (UL) and lower-right (LR) corners of the box, etc. Each run-length has a pointer to the next run-length of the same color (null if none exists) and an index corresponding to the bounding box that it belongs to, while each bounding box has a pointer to the list of run-lengths that it encloses. This facilitates the easy merging of two run-lengths (or a bounding box containing several run-lengths with a single run-length or two bounding boxes each having more that one run-length). The RunRegion data structure and the BoundingBox data structure are given in Table 4.1.

TABLE 4.1: This Table Shows the Basic RunRegion and BoundingBox Data Structures With Which We Operate

```
// The Runregion data structure definition.

struct RunRegion {
    int color;   //color associated with the run region.
    RunRegion* root;  //the root node of the runregion.
    RunRegion* listNext;  //pointer to the next runregion in the current run length.
    RunRegion* nextRun;
    int xLoc;  //x location of the root node.
    int yLoc;  //y location of the root node.
    int runLength;  // number of run lengths with this region.
    int boundingBox;  //the bounding box that this region belongs to.
};


// The  BoundingBox  data structure definition.

struct BoundingBox {
    BoundingBox*  prevBox; //pointer to the previous bounding box.
    BoundingBox*  nextBox;  // pointer to the next bounding box.
    int  ULx; //upper left corner x coordinate.
    int  ULy; //upper left corner y coordinate.
    int  LRx;
    int  LRy;
    bool  lastBox;
    int  valid;
    int  numRunLengths; //number of runlengths associated with this bounding box.
    int  numPixels;  //number of pixels in this bounding box.
    int  rrcount;
    int  color; //color cooresponding to this bounding box.
    RunRegion*  listRR;
    RunRegion*  eoList;
    float  prob;  //probability corresponding to this bounding box.
};
```

Next, we need to merge the run-lengths/bounding boxes corresponding to the same object together under the assumption that an object in the image will be represented by connected run-lengths (see [58] for a description of some techniques for performing the merging). In the second pass, we proceed along the run-lengths (in the order in which they are present in the linked list) and check for pixels of the same color immediately below each pixel over which the run-length extends, merging run-lengths of the same color that have significant overlap (the threshold number of pixel overlap is decided based on experimentation: see Appendix A.1). When two run-lengths are to be merged, one of the bounding boxes is deleted while the other's properties (root node, number of run-lengths, size, etc.) are suitably modified to include both the bounding boxes. This is accomplished by moving the corresponding pointers around appropriately. By incorporating suitable heuristics, we remove bounding boxes that are not significantly large or dense enough to represent an object of interest in the image, and at the end of this pass, we end up with a number of candidate bounding boxes, each representing a blob of one of the nine colors under consideration. The bounding boxes corresponding to each color are linked together in a separate linked list, which (if required) is sorted in descending order of size for ease of further processing. Details of the heuristics used here can be found in Appendix A.1.

When processing images in real time, the low-level vision components described in Sections 4.2 and 4.3—color segmentation and blob formation—take ≈20 ms per frame (of the available ≈33 ms).

4.4   OBJECT RECOGNITION WITH BOUNDING BOXES

Once we have bounding boxes of the various colors arranged in separate lists (blobs), we can proceed to high-level vision, i.e., the detection of objects of interest in the robot's image frame. Object recognition is a major area of research in computer vision and several different approaches have been presented, depending on the application domain [5, 68, 88]. Most of these approaches either involve extensive computation of object features or large amounts of storage in the form of object templates corresponding to different views, making them infeasible in our domain. Further they are not very effective for rapidly changing camera positions. We determine the objects of interest in the image using domain knowledge rather than trying to extract additional features from the image.

The objects that we primarily need to identify in the visual field are the ball, the two goals, the field markers (other than the goals), and the opponents. This stage takes as input the lists of bounding boxes and provides as output a collection of objects (structures called the *VisionObjects*), one for each detected object, which are then used for determining the position and bearing of these objects with respect to the robot. This information is in turn used in the localization module (see Chapter 8) to calculate the robot's position in the field coordinates. To identify these objects, we introduce some constraints and heuristics, some of which are based on the known geometry of the environment while others are parameters that we identified by the experimentation. First, the basic process used to search for the various objects is documented, and at the end of the chapter, a description of the constraints and heuristics used is provided.

We start with the goals because they are generally the largest blobs of the corresponding colors and once found they can be used to eliminate spurious blobs during beacon and ball detection. We search through the lists of bounding boxes for colors corresponding to the goals (blue and yellow) on the field, given constraints on size, aspect ratio, and density Furthermore, checks are included to ensure that spurious blobs (noisy estimates on the field, blobs floating in the air, etc.) are not taken into consideration. On the basis of these constraints, we compare the blob found in the image (and identified as a goal) with the known geometry of the goal. This provides some sort of likelihood measure, and a *VisionObject* is created to store this and the information of the corresponding bounding box. (Table 4.2 displays the data structures used for this purpose.)

After searching for the goals, we search for the orange ball, probably the most important object in the field. We sort the orange bounding boxes in descending order of size and search

TABLE 4.2: This Table Shows the Basic Visionobject and Associated Data Structures With Which We Operate

```
struct VisionObjects{
    int NumberOfObjects;  //number of vision obejcts in curretn frame.
    BBox*  ObjectInfo;    //array of objects in view.
}

struct BBox {
   int ObjID; //object ID.
   Point ul;  //upper left point of the bounding box.
   Point lr;  //lower right point of the bounding box.
   double  prob; //likelihood corresponding to this bounding box/object.
}

struct Point {
    double x;  //x coordinate.
    double  y;  //y coordinate.
}
```

through the list (not considering very small ones), once again based on heuristics on size, aspect ratio, density, etc. To deal with cases with partial occlusions, which is quite common with the ball on the field, we use the "circle method" to estimate the equation of the circle that best describes the ball (see Appendix A.3 for details). Basically, this involves finding three points on the edge of the ball and finding the equation of the circle passing through the three points. This method seems to give us an accurate estimate of the ball size (and hence the ball distance) in most cases. In the case of the ball, in addition to the check that helps eliminate spurious blobs (floating in the air), checks have to be incorporated to ensure that minor misclassification in the segmentation stage (explained below) do not lead to detection of the ball in places where it does not exist.

Next, we tackle the problem of finding the beacons (six field markers, excluding the goals). The identification of beacons is important in that the accuracy of localization of the robot depends on the determination of the position and bearing of the beacons (with respect to the robots) which in turn depends on the proper determination of the bounding boxes associated with the beacons. Since the color pink appears in all beacons, we use that as the focus of our search. Using suitable heuristics to account for size, aspect ratio, density, etc., we match each pink blob with blue, green, or yellow blobs to determine the beacons. We ensure that only one instance of each beacon (the most likely one) is retained. Additional tests are incorporated to remove spurious beacons: those that appear to be on the field or in the opponents, floating in the air, inappropriately huge or tiny, etc. For details, see Appendix A.4.

After this first pass, if the goals have not been found, we search through the candidate blobs of the appropriate colors with a set of reduced constraints to determine the occurrence of the goals (which results in a reduced likelihood estimate as we will see below). This is useful when we need to identify the goals at a distance, which helps us localize better, as each edge of the goal serves as an additional marker for the purpose of localization.

We found that the goal edges were much more reliable as inputs to the localization module than were the goal centers. So, once the goals are detected, we determine the edges of the goal based on the edges of the corresponding bounding boxes. Of course, we include proper buffers at the extremities of the image to ensure that we detect the actual goal edges and not the 'artificial edges' generated when the robot is able to see only a section of the goal (as a result of its view angle) and the sides of the truncated goal's bounding box are mistaken to be the actual edges.

Next, a brief description of some of the heuristics employed in the detection of the ball, goals, beacons, and opponents is presented. I begin by listing the heuristics that are common to all objects and then also list those that are specific to goals, ball, and/or beacons. For more detailed explanations on some methods and parameters for individual test, see the corresponding appendices.

- *Spurious blob elimination:* A simple calculation using the tilt angle of the robot's head is used to determine and hence eliminate spurious (beacon, ball, and/or goal) blobs that are too far down or too high up in the image plane. See Appendix A.2 for the actual thresholds and calculations.

- *Likelihood calculation:* For each object of interest in the robot's visual field, we associate a measure which describes how sure we are of our estimation of the presence of that object in the current image frame. The easiest way to accomplish this would be to compare the aspect ratio (the ratio of the height to the width) of the bounding boxes that identify these objects, to the actual known aspect ratio of the objects in the field. For example, the goal has an aspect ratio of 1 : 2 in the field, and we can compare the aspect ratio of the bounding box that has been detected as the goal with this expected ratio. We can claim that the closer these two values are, the more sure we are of our estimate and hence higher is the *likelihood*.

- *Beacon-specific calculations:*
  (1) To remove spurious beacons, we ensure that the two sections that form the beacon are of comparable size, i.e., each section is at least half as large and half as dense as the other section.

  (2) We ensure that the separation between the two sections is within a small threshold, which is usually 2–3 pixels.

  (3) We compare the aspect ratio of bounding box corresponding to the beacon in the image to the actual aspect ratio (2:1 :: *height* : *width*), which helps remove candidate beacons that are too small or disproportionately large.

  (4) Aspect ratio, as mentioned above, is further used to determine an estimate of the likelihood of each candidate beacon that also helps to choose the "most probable" candidate when there are multiple occurrences of the same beacon. Only beacons with a likelihood above a threshold are retained and used for localization calculations. This helps to ensure that false positives, generated by lighting variations and/or shadows, do not cause major problems in the localization.

  Note: For sample threshold values, see Appendix A.4.

- *Goal-specific calculations:*
  (1) We use the 'tilt-angle test' (described in detail in Appendix A.2).

  (2) We use a similar aspect ratio test for the goals, too. In the case of the goals, we also look for sufficiently high density (the ratio of the number of pixels of the

expected color to the area of the blob), the number of run-lengths enclosed, and a minimum number of pixels. All these thresholds were determined experimentally, and changing these thresholds changes the distance from which the goal can be detected and the accuracy of detection. For example, lowering these thresholds can lead to false positives.

(3) The aspect ratio is used to determine the likelihood, and the candidate is accepted if and only if it has a likelihood measure above a predefined minimum.

(4) When doing a second pass for the goal search, we relax the constraints slightly but proportionately a lower likelihood measure gets assigned to the goal, if detected.

Note: For sample threshold values, see Appendix A.5.

- *Ball-specific calculations:*
(1) We use the 'tilt-angle test' to eliminate spurious blobs from consideration.

(2) In most cases, the ball is severely occluded, precluding the use of the aspect ratio test. Nonetheless, we first search for an orange object with a high density and an aspect ratio (1:1) that would detect the ball if it is seen completely and not occluded.

(3) If the ball is not found with these tight constraints, we relax the aspect ratio constraint and include additional heuristics (e.g., if the ball is close, even if it is partially occluded, it should have a large number of run-lengths and pixels) that help detect a bounding box around the partially occluded ball. These heuristics and associated thresholds were determined experimentally.

(4) If the yellow goal is found, we ensure that the candidate orange ball does not occur within it and above the ground (which can happen since yellow and orange are close in color space).

(5) We check to make sure that the orange ball is found lower than the lower-most beacon in the current frame. Also, the ball cannot occur above the ground, or within or slightly below the beacon. The latter can occur if the white and/or yellow portions of the beacon are misclassified as orange.

(6) We use the "circle method" to detect the actual ball size. But we also include checks to ensure that in cases where this method fails and we end up with disproportionately huge or very small ball estimates (thresholds determined experimentally), we either keep the estimates we had before employing the circle method (and extend the bounding box along the shorter side to form a square to get the closest approximation to the ball) or reject the ball estimate in the current frame. The

choice depends on the extent to which the estimated "ball" satisfies experimental thresholds.

Note: For sample threshold values, see Appendix A.6.

Finally, we check for opponents in the current image frame. As in the previous cases, suitable heuristics are employed both to weed out the spurious cases and to determine the likelihood of the estimate. To identify the opponents, we first sort the blobs of the corresponding color in descending order of size, with a minimum threshold on the number of pixels and run-lengths. We include a relaxed version of the aspect ratio test and strict tilt-angle tests (an "opponent" blob cannot occur much lower or much higher than the horizon when the robot's head has very little tilt and roll) to further remove spurious blobs (see Appendix A.2 and Appendix A.7). Each time an opponent blob (that satisfies these thresholds) is detected, the robot tries to merge it with one of its previous estimates based on thresholds. If it does not succeed and it has less than four valid (previous) estimates, it adds this estimate to the list of opponents. At the end of this process, each robot has a list that stores the four largest bounding boxes (that satisfy all these tests) of the color of the opponent with suitable likelihood estimates that are determined based on the size of the bounding boxes (see Appendix A.8). Further processing of opponent estimates using the estimates from other teammates, etc., is described in detail in the section on visual opponent modeling (Section 4.6). Once the processing of the current visual frame is completed, the detected objects, each stored as a VisionObject are passed through the *Brain* (central control module as described in Chapter 10) to the GlobalMap module wherein the VisionObjects are operated upon using Localization routines.

4.5   POSITION AND BEARING OF OBJECTS

The object recognition module returns a set of data structures, one for each "legal" object in the visual frame. Each object also has an estimate of its likelihood, which represents the uncertainty in our perception of the object. The next step (the final step in high-level vision) is to determine the distance to each such object from the robot and the bearing of the object with respect to the robot. In our implementation, this estimation of distance and bearing of all objects in the image, with respect to the robot, is done as a preprocessing step when the localization module kicks into action during the development of the global maps. Since this is basically a vision-based process, it is described here rather than in the chapter (Chapter 8) on localization. As each frame of visual input is processed, the robot has access to the tilt, pan, and roll angles of its camera from the appropriate sensors and these values give us a simple transform that takes us from the 3D world to the 2D image frame. Using the known projection of the object in the image plane and the geometry of the environment (the expected sizes of the objects in the robot's environment), we can arrive at estimates for the distance and bearing of the object relative to the robot. The known geometry is used to arrive at an estimate for the variances corresponding to the distance and the bearing. Suppose the distance and angle estimates for a beacon are d and θ. Then the variances in the distance and bearing estimates are estimated as

$$\text{variance}_d = \left(\frac{1}{b_p}\right) \cdot (0.1d) \qquad (4.4)$$

where $\left(\frac{1}{b_p}\right)$ is the likelihood of the object returned by vision.

$$\text{variance}_\theta = \tan^{-1}\left(\frac{\text{beacon}_r}{d}\right) \qquad (4.5)$$

where beacon_r is the actual radius of the beacon in the environment.

By similar calculations, we can determine the distance and bearing (and the corresponding variances) of the various objects in the robot's field of view.

4.6   VISUAL OPPONENT MODELING

Another important task accomplished using the image data is that of opponent modeling. As described in Section 4.4, each robot provides a maximum of four best estimates of the opponent blobs based on the current image frame. To arrive at an efficient estimate of the opponents (location of the opponents relative to the robot and hence with respect to the global frame), each robot needs to merge its own estimates with those communicated by its teammates. As such this process is accomplished during the development of the global maps (Chapter 11) but since the operation interfaces directly with the output from the vision module, it is described here.

When opponent blobs are identified in the image frame, the vision module returns the bounding boxes corresponding to these blobs. We noticed that though the shape of the blob and hence the size of the bounding box can vary depending on the angle at which the opponent robot is viewed (and its relative orientation), the height of the bounding box is mostly within a certain range. We use this information to arrive at an estimate of the distance of the opponent and use the centroid of the bounding box to estimate the bearing of the candidate opponent with respect to the robot (see Section 4.5 for details on estimation of distance and bearing of objects). These values are used to find the opponent's (x, y) position relative to the robot and hence determine the opponent's global (x, y) position (see Appendix A.9 for details on transforms from local to global coordinates and vice versa). Variance estimates for both the x and the y positions are obtained based on the calculated distance and the likelihood associated with that particular opponent blob. For example, let d and θ be the distance and bearing of the opponent relative to the robot. Then, in the robot's local coordinate frame (determined by the robot's position and orientation), we have the relative positions as

$$\text{rel}_x = d \cdot \cos(\theta), \qquad \text{rel}_y = d \cdot \sin(\theta).$$

From these, we obtain the global positions as

$$\begin{pmatrix} \text{glob}_x \\ \text{glob}_y \end{pmatrix} = T_{\text{local}}^{\text{global}} \cdot \begin{pmatrix} \text{rel}_x \\ \text{rel}_y \end{pmatrix} \qquad (4.6)$$

where $T_{\text{local}}^{\text{global}}$ is the 2D-transformation matrix from local to global coordinates.

For the variances in the positions, we use a simple approach:

$$\text{var}_x = \text{var}_y = \frac{1}{\text{Opp}_{\text{prob}}} \cdot (0.1d) \qquad (4.7)$$

where the likelihood of the opponent blob, Opp_{prob} is determined by heuristics (see Appendix A.8).

If we do not have any previous estimates of opponents from this or any previous frame, we accept this estimate and store it in the list of known opponent positions. If any previous estimates exist, we try to merge them with the present estimate by checking if they are close enough (based on heuristics). All merging is performed assuming Gaussian distributions. The basic idea is to consider the x and y position as independent Gaussians (with the positions as the means and the associated variances) and merge them (for more details, see [84]). If merging is not possible and we have fewer than four opponent estimates, we treat this as a new opponent estimate and store it as such in the opponents list. But if four opponent estimates already exist, we try to replace one of the previous estimates (the one with the maximum variance in the list of opponent estimates and with a variance higher than the new estimate) with the new estimate. Once we have traversed through the entire list of opponent bounding boxes presented by the vision module, we go through our current list of opponent estimates and degrade all those estimates that were not updated, i.e., not involved in merging with any of the estimates from the current frame (for more details on the degradation of estimates, see the initial portions of Chapter 11 on global maps). When each robot shares its Global Map (see Chapter 11) with its teammates, these data get communicated.

When the robot receives data from its teammates, a similar process is incorporated. The robot takes each current estimate (i.e., one that was updated in the current cycle) that is communicated by a teammate and tries to merge it with one of its own estimates. If it fails to do so and it has fewer than four opponent estimates, it accepts the communicated estimate as such and adds it to its own list of opponent estimates. But if it already has four opponent estimates, it replaces its oldest estimate (the one with the largest variance which is larger than the variance of the communicated estimate too) with the communicated estimate. If this is not possible, the communicated estimate is ignored.

This procedure, though simple, gives reliable results in nearly all situations once the degradation and merging thresholds are properly tuned. It was used both during games and in one of the challenge tasks (see Appendix E.3) during RoboCup and the performance was good enough to walk from one goal to the other avoiding all seven robots placed in its path.

The complete vision module as described in this chapter, starting from color segmentation up to and including the object recognition phase takes ≈28 ms per frame, enabling us to process images at frame rate. Though the object recognition algorithm described here does not recognize lines in the environment, they are also a great source of information, and can be detected reliably with a time cost of an additional ≈3 ms per frame, thus still staying within frame rate [73].

CHAPTER 5

Movement

Enabling a robot to move precisely and quickly is equally as essential to any interesting task, including the RoboCup task, as is vision. In this chapter, our approach to AIBO movement is introduced, including walking and the interfaces from walking to the higher level control modules.

Just as vision-based robots are increasingly replacing robots with primarily range sensors, it is now becoming possible to deploy legged robots rather than just wheeled robots. The insights from this section are particularly relevant to such legged robotics.

The AIBO comes with a stable but slow walk. From watching the videos of past RoboCups, and from reading the available technical reports, it became clear that a fast walk is an essential part of any RoboCup team. The walk is perhaps the most feasible component to borrow from another team's code base, since it can be separated out into its own module. Nonetheless, we decided to create our own walk in the hopes of ending up with something at least as good, if not better, than that of other teams, while retaining the ability to fine tune it on our own.

The movement component of our team can be separated into two parts. First, the walking motion itself, and second, an interface module between the low-level control of the joints (including both walking and kicking) and the decision-making components.

5.1 WALKING

This section details our approach to enabling the AIBOs to walk. Though it includes some AIBO-specifics, it is for the most part a general approach that could be applied to other legged robots with multiple controllable joints in each leg.

5.1.1 Basics

At the lowest level, walking is effected on the AIBO by controlling the joint angles of the legs. Each of the four legs has three joints known as the rotator, abductor, and knee. The rotator is a shoulder joint that rotates the entire leg (including the other two joints) around an axis that runs horizontally from left to right. The abductor is the shoulder joint responsible for rotating the leg out from the body. Finally, the knee allows the lower link of the leg to bend forwards or backwards, although the knees on the front legs primarily bend the feet forwards while the ones on the back legs bend primarily backwards. These rotations will be described more precisely in the section on forward kinematics.

Each joint is controlled by a PID mechanism. This mechanism takes as its inputs P, I, and D gain settings for that joint and a desired angle for it. The robot architecture can process a request for each of the joints at a rate of at most once every 8 ms. We have always requested joint values at this maximum allowed frequency. Also, the AIBO model information lists recommended settings for the P, I, and D gains for each joint. We have not thoroughly experimented with any settings aside from the recommended ones and use only the recommended ones for everything that is reported here.

The problem of compelling the robot to walk is greatly simplified by a technique called inverse kinematics. This technique allows the trajectory of a leg to be specified in terms of a three-dimensional trajectory for the foot. The inverse kinematics converts the location of the foot into the corresponding settings for the three joint angles. A precursor to deriving inverse kinematics formulas is having a model of the forward kinematics, the function that takes the three joint angles to a three-dimensional foot position. This is effectively our mathematical model of the leg.

5.1.2   Forward Kinematics

For each leg, we define a three-dimensional coordinate system whose origin is at the leg's shoulder. In these coordinate systems, positive x is to the robot's right, positive y is the forward direction, and positive z is up. Thus, when a positive angle is requested from a certain type of joint, the direction of the resulting rotation may vary from leg to leg. For example, a positive angle for the abductor of a right leg rotates the leg out from the body to the right, while a positive angle for a left leg rotates the leg out to the left. We will describe the forward and inverse kinematics for the front right leg, but because of the symmetry of the AIBO, the inverse kinematics formulas for the other legs can be attained simply by first negating x or y as necessary.

The unit of distance in our coordinate system is the length of one link of any leg, i.e. from the shoulder to the knee, or from the knee to the foot. This may seem a strange statement, given that, physically speaking, the different links of the robot's legs are not exactly the same length. However, in our mathematical model of the robot, the links are all the same length. This serves to simplify our calculations, although it is admittedly an inaccuracy in our model. We argue that this inaccuracy is overshadowed by the fact that we are not modeling the leg's foot, a cumbersome unactuated aesthetic appendage. As far as we know, no team has yet tried to model the foot.

We call the rotator, abductor, and knee angles J_1, J_2, and J_3, respectively. The goal of the forward kinematics is to define the function from $J = (J_1, J_2, J_3)$ to $p = (x, y, z)$, where p is the location of the foot according to our model. We call this function $K_F(J)$. We start with the fact that when $J = (0, 0, 0)$, $K_F(J) = (0, 0, -2)$, which we call p_0. This corresponds to the situation where the leg is extended straight down. In this base position for the leg, the knee is at the point $(0, 0, -1)$. We will describe the final location of the foot as the result of a series of three rotations being applied to this base position, one for each joint.

First, we associate each joint with the rotation it performs when the leg is in the base position. The rotation associated with the knee, $K(q, \Theta)$, where q is any point in space, is a rotation around the line $y = 0$, $z = -1$, counterclockwise through an angle of Θ with the x-axis pointing towards you. The abductor's rotation, $A(q, \Theta)$, goes clockwise around the y-axis. Finally, the rotator is $R(q, \Theta)$, and it rotates counterclockwise around the x-axis. In general (i.e. when J_1 and J_2 are not 0), changes in J_2 or J_3 do not affect p by performing the corresponding rotation A or K on it. However, these rotations are very useful because the forward kinematics function can be defined as

$$K_F(J) = R(A(K(p_0, J_3), J_2), J_1). \tag{5.1}$$

This formulation is based on the idea that for any set of angles J, the foot can be moved from p_0 to its final position by rotating the knee, abductor, and rotator by J_3, J_2, and J_1, respectively, *in that order*. This formulation works because when the rotations are done in that order they are always the rotations K, A, and R. A schematic diagram of the AIBO after each of the first two rotations is shown in Fig. 5.1.

It is never necessary for the robot to calculate x, y, and z from the joint angles, so the above equation need not be implemented on the AIBO. However, it is the starting point for the derivation of the Inverse Kinematics, which are constantly being computed while the AIBO is walking.

5.1.3   Inverse Kinematics

Inverse kinematics is the problem of finding the inverse of the forward kinematics function K_F, $K_I(q)$. With our model of the leg as described above, the derivation of K_I can be done by a relatively simple combination of geometric analysis and variable elimination.

The angle J_3 can be determined as follows. First, we calculate d, the distance from the shoulder to the foot, which is given by

$$d = \sqrt{x^2 + y^2 + z^2}. \tag{5.2}$$

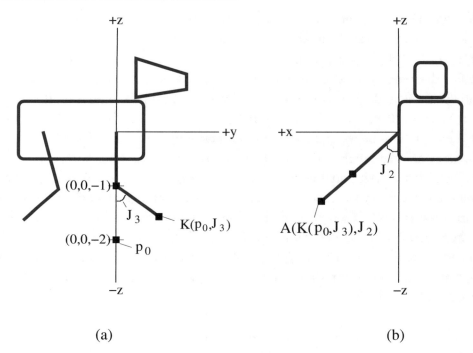

FIGURE 5.1: Schematic drawings of the AIBO according to our kinematics model. (a) This is a side view of the AIBO after rotation K has been performed on the foot. (b) In this front view, rotation A has also been performed.

Next, note that the shoulder, knee, and foot are the vertices of an isosceles triangle with sides of length 1, 1, and d with central angle $180 - J_3$. This yields the formula

$$J_3 = 2\cos^{-1}\left(\frac{d}{2}\right). \tag{5.3}$$

The inverse cosine here may have two possible values within the range for J_3. In this case, we always choose the positive one. While there are some points in three-dimensional space that this excludes (because of the joint ranges for the other joints), those points are not needed for walking. Furthermore, if we allowed J_3 to sometimes be negative, it would be very difficult for our function K_I to be continuous over its entire domain.

To compute J_2, we must first write out an expression for $K(p_0, J_3)$. It is $(0, \sin J_3, 1 + \cos J_3)$. This is the position of the foot in Fig. 5.1(a). Then we can isolate the effect of J_2 as follows. Since the rotation R is with respect to the x-axis, it does not affect the x-coordinate. Thus, we can make use of the fact that the $K_F(J)$, which is defined to be $R(A(K(p_0, J_3), J_2), J_1)$ (Eq. (5.1)), has the same x-coordinate as $A(K(p_0, J_3), J_2)$. Plugging in our expression for

$K(p_0, J_3)$, we get that

$$A(K(p_0, J_3), J_2) = A((0, \sin J_3, 1 + \cos J_3), J_2). \tag{5.4}$$

Since A is a rotation around the y-axis,

$$A(K(p_0, J_3), J_2) = (\sin J_2(1 + \cos J_3), \sin J_3, \cos J_2(1 + \cos J_3)). \tag{5.5}$$

Setting x (which is defined to be the x-coordinate of $K_F(J)$) equal to the x-coordinate here and solving for J_2 gives us

$$J_2 = \sin^{-1}\left(\frac{x}{1 + \cos J_3}\right). \tag{5.6}$$

Note that this is only possible if $x \le 1 + \cos(J_3)$. Otherwise, there is no J_2 that satisfies our constraint for it, and, in turn, no J such that $F_K(J) = q$. This is the *impossible sphere* problem, which we discuss in more detail below. The position of the foot after rotations K and A is depicted in Fig. 5.1(b).

Finally, we can calculate J_1. Since we know y and z before and after the rotation R, we can use the difference between the angles in the y-z plane of the two (y, z)'s. The C++ function atan2(z, y) gives us the angle of the point (y, z), so we can compute

$$J_1 = \text{atan2}(z, y) - \text{atan2}(\cos J_2(1 + \cos J_3), \sin J_3). \tag{5.7}$$

The result of this subtraction is normalized to be within the range for J_1. This concludes the derivation of J_1 through J_3 from x, y, and z. The computation itself consists simply of the calculations in the four equations (5.2), (5.3), (5.6), and (5.7).

It is worth noting that expressions for J_1, J_2, and J_3 are never given explicitly in terms of x, y, and z. Such expressions would be very convoluted, and they are unnecessary because the serial computation given here can be used instead. Furthermore, we feel that this method yields some insight into the relationships between the legs joint angles and the foot's three-dimensional coordinates.

There are many points q, in three-dimensional space, for which there are no joint angles J such that $F_K(J) = q$. For these points, the inverse kinematics formulas are not applicable. One category of such points is intuitively clear: the points whose distance from the origin is greater than 2. These are impossible locations for the foot because the leg is not long enough to reach them from the shoulder. There are also many regions of space that are excluded by the angle ranges of the three joints. However, there is one unintuitive, but important, unreachable region, which we call the impossible sphere. The impossible sphere has a radius of 1 and is centered at the point $(1, 0, 0)$. The following analysis explains why it is impossible for the foot to be in the interior of this sphere.

Consider a point (x, y, z) in the interior of the illegal sphere. This means that

$$(x - 1)^2 + y^2 + z^2 < 1$$
$$x^2 - 2x + 1 + y^2 + z^2 < 1$$
$$x^2 + y^2 + z^2 < 2x.$$

Substituting d for $\sqrt{x^2 + y^2 + z^2}$ and dividing by 2 gives us

$$\frac{d^2}{2} < x. \qquad (5.8)$$

Since $J_3 = 2\cos^{-1}\left(\frac{d}{2}\right)$ (Eq. (5.3)), $\cos\frac{J_3}{2} = \frac{d}{2}$, so by the double-angle formula, $\cos J_3 = \frac{d^2}{2} - 1$, or $\frac{d^2}{2} = 1 + \cos J_3$. Substituting for $\frac{d^2}{2}$, we get

$$x > 1 + \cos J_3. \qquad (5.9)$$

This is precisely the condition, as discussed above, under which the calculation of J_2 breaks down. This shows that points in the illegal sphere are not in the range of F_K.

Occasionally, our parameterized walking algorithm requests a position for the foot that is inside the impossible sphere. When this happens, we project the point outward from the center of the sphere onto its surface. The new point on the surface of the sphere is attainable, so the inverse kinematics formulas are applied to this point.

5.1.4   General Walking Structure

Our walk uses a trot-like gait in which diagonally opposite legs step together. That is, first one pair of diagonally opposite legs steps forward while the other pair is stationary on the ground. Then the pairs reverse roles so that the first pair of legs is planted while the other one steps forward. As the AIBO walks forward, the two pairs of diagonally opposite legs continue to alternate between being on the ground and being in the air. For a brief period of time at the start of our developmental process, we explored the possibility of other gait patterns, such as a walking gait where the legs step one at a time. We settled on the trot gait after watching video of RoboCup teams from previous years.

While the AIBO is walking forwards, if two feet are to be stationary on the ground, that means that they have to move backwards with respect to the AIBO. In order for the AIBO's body to move forwards in a straight line, each foot should move backwards in a straight line for this portion of its trajectory. For the remainder of its trajectory, the foot must move forward in a curve through the air. We opted to use a half-ellipse for the shape of this curve (Fig. 5.2).

A foot's half-elliptical path through the air is governed by two functions, $y(t)$ and $z(t)$, where t is the amount of time that the foot has been in the air so far divided by the total time

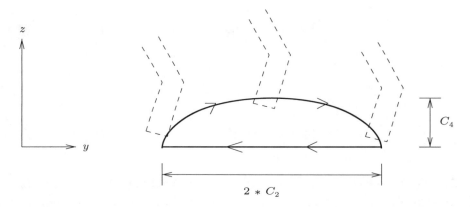

FIGURE 5.2: The foot traces a half-ellipse as the robot walks forward.

the foot spends in the air (so that t runs from 0 to 1). While the AIBO is walking forwards, the value of x for any given leg is always constant. The values of y and z are given by

$$y(t) = C_1 - C_2 \cos(\pi t) \tag{5.10}$$

and

$$z(t) = C_3 - C_4 \sin(\pi t). \tag{5.11}$$

In these equations, C_1 through C_4 are four parameters that are fixed during the walk. C_1 determines how far forward the foot is and C_3 determines how close the shoulder is to the ground. The parameters C_2 and C_4 determine how big a step is and how high the foot is raised for each step (Fig. 5.2). Our walk has many other free parameters, which are all described in Section A.9.1.

5.1.5   Omnidirectional Control

After implementing the forward walk, we needed sideways, backwards, and turning motions. There is a nice description of how to obtain all these (and any combination of these types of walks) in [32]. We based our implementation on the ideas from that paper.

Sideways and backwards walks are just like the forward walk with the ellipse rotated around the z-axis (Fig. 5.3(a)). For walking sideways, the ellipse is rotated 90° to the side towards which the robot should walk. For walking backwards, the ellipse points in the negative y-direction. Turning in place is a little more complicated. The four legs of the robot define a circle passing through them. The direction of the ellipse for each leg is tangent to this circle, pointing clockwise if the robot is to turn right and counterclockwise to turn left (Fig. 5.3(b)).

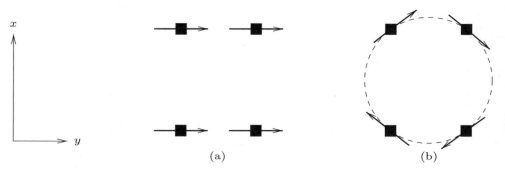

(a)   (b)

FIGURE 5.3: The main movement direction of the half-ellipses changes for (a) walking sideways and (b) turning in place. (The dark squares indicate the positions of the four feet when standing still.)

Combinations of walking forwards, backwards, sideways, and turning are also possible by simply combining the different components for the ellipses through vector addition. For example, to walk forwards and to the right at the same time, at an angle of $45°$ to the y-axis, we would make the ellipses point $45°$ to the right of the y-axis. Any combination can be achieved as shown in Fig. 5.4.

In practice, the method described here worked well for combinations of forwards and turning velocities, but we had difficulty also incorporating sideways velocities. The problem was that, after tuning the parameters (Section 5.1.7), we found that the parameters that worked well for going forwards and turning did not work well for walking sideways. It was not obvious how to find common parameters that would work for combinations of all the three types of velocities.

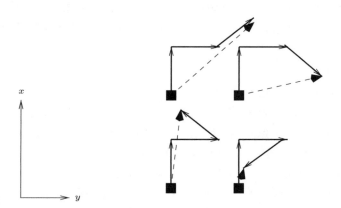

FIGURE 5.4: Combining forwards, sideways, and turning motions. Each component contributes a vector to the combination. Dashed lines show the resulting vectors. (We show only half of the ellipse lengths, for clarity.) With the vectors shown, the robot will be turning towards its right as it moves diagonally forward and right.

In situations where we needed to walk with a nonzero sideways velocity, we frequently used a slower omnidirectional walk developed by a student in the Spring semester class.[1] That walk is called SPLINE_WALK, while the one being described here is called PARAM_WALK. Section 5.2.3 discusses when each of the walks was used.

More recent walking engines by some other teams take a similar approach to the ones described in this section, but are somewhat more flexible with regards to providing smooth omnidirectional motion and a set of different locus representations including rectangular, elliptic, and Hermite curve shaped trajectories [2]. Although, those representations provide reasonable results, the biggest constraint they have in common are that they are 2D representations. The German team showed that representing loci in 3D yields a better walk both in terms of speed and in maneuverability [20].

5.1.6   Tilting the Body Forward

Up until the 2003 American Open, our walking module was restricted to having the AIBO's body be parallel to the ground. That is, it did not allow for the front and back shoulders to be different distances from the ground. This turned out to be a severe limitation. During this time, we were unable to achieve a forward speed of over 150 mm/s. After relaxing this constraint, only the slightest hand tuning was necessary to bring our speed over 200 mm/s. After a significant amount of hand tuning, we were able to achieve a forward walking speed of 235 mm/s. (The parameters that achieve this speed are given in Section 5.1.7 and our procedure for measuring walking speed is described in Section 5.1.8.)

In many of the fastest and most stable walks, the front legs touch the ground with their elbows when they step. Apparently, this is far more effective than just having the feet touch the ground. We enable the elbows to touch the ground by setting the height of the front shoulders to be lower than that of the back shoulders. However, this ability requires one more computation to be performed on the foot coordinates before the inverse kinematics equations are applied. That is, when the AIBO's body is tilted forward, we still want the feet to move in half-ellipses that run parallel to the ground. This means that the points given by Eqs. (5.10) and (5.11) have to be rotated with respect to the x-axis before the inverse kinematics equations are applied.

The angle through which these points must be rotated is determined by the difference between the desired heights of the front and back shoulders and the distance between the front and back shoulders. The difference between the heights, d_h, is a function of the parameters being used (the heights of the front and back shoulders are two of our parameters), but the distance between the front and back shoulders is a fixed body length distance which we estimate

---

[1]Aniket Murarka.

at 1.64 in our units and call l_b. Then, the angle of the body rotation is given by

$$\theta = \sin^{-1}\left(\frac{d_b}{l_b}\right). \tag{5.12}$$

5.1.7   Tuning the Parameters

Once the general framework of our walk was set up, we were faced with the problem of determining good values for all of the parameters of the walk. The complete set of parameters resulting from our elliptical locus is listed in Appendix A.9.1.

Eventually, we adopted a machine learning approach to this problem (see Section 5.3.1). But initially, the tuning process was greatly facilitated by the use of a tool we had written that allowed us to telnet into the AIBO and change the walking parameters at run time. Thus, we were able to go back and forth between altering parameters and watching (or timing) the AIBO to see how fast it was. This process enabled us to experiment with many different combinations of parameters.

We focused most of our tuning effort on finding as fast a straight-forward walk as possible. Our tuning process consisted of a mixture of manual hill-climbing and using our observations of the walk and intuition about the effects of the parameters. For example, two parameters that were tuned by relatively blind hill-climbing were Forward step distance and Moving_max_counter. These parameters are very important and it is often difficult to know intuitively if they should be increased or decreased. So tuning proceeded slowly and with many trials. On the other hand, parameters such as the front and back clearances could frequently be tuned by noticing, for instance, that the front (or back) legs dragged along the ground (or went too high in the air).

5.1.8   Odometry Calibration

As the AIBO walks, it keeps track of its forward, horizontal, and angular velocities. These values are used as inputs to our particle filtering algorithm (see Chapter 8) and it is important for them to be as accurate as possible. The Movement Module takes a walking request in the form of a set of forward, horizontal, and angular velocities. These velocities are then converted to walking parameters. The "Brain" (Chapter 10) assumes that the velocities being requested are the ones that are actually attained, so the accuracy of the odometry relies on that of those conversions.

Since the step distance parameters are proportional to the distance traveled each step and the time for each step is the same, the step distance parameters should theoretically be proportional to the corresponding velocities. This turned out to be true to a fair degree of accuracy for combinations of forward and turning velocities. As mentioned above, we needed to use a different set of parameters for walking with a nonzero sideways velocity. These parameters did not allow for a fast forward walk, but with them the velocities were roughly proportional to the step distances for combinations of forward, turning, and sideways velocities.

The proportionality constants are determined by a direct measurement of the relevant velocities. To measure forward velocity, we use a stopwatch to record the time the robot takes to walk from one goal line to the other with its forward walking parameters. The time taken is divided into the length of the field, 4200 mm, to yield the forward velocity. The same process is used to measure sideways velocity. To measure angular velocity, we execute the walk with turning parameters. Then we measure how much time it takes to make a certain number of complete revolutions. This yields a velocity in degrees per second. Finally, the proportionality constants were calculated by dividing the measured velocities by the corresponding step distance parameters that gave rise to them.

Since the odometry estimates are used by localization (Chapter 8), the odometry calibration constants could be tuned more precisely by running localization with a given set of odometry constants and observing the effects of the odometry on the localization estimates. Then we could adjust the odometry constants in the appropriate direction to make localization more accurate. We feel that we were able to achieve quite accurate odometry estimates by a repetition of this process.

5.2   GENERAL MOVEMENT

Control of the AIBO's movements occurs at three levels of abstraction.

1. The lowest level, the "movement module," resides in a separate Open-R object from the rest of our code (as described in the context of our general architecture in Chapter 10) and is responsible for sending the joint values to *OVirtualRobotComm*, the provided Open-R object that serves as an interface to the AIBO's motors.

2. One level above the movement module is the "movement interface," which handles the work of calculating many of the parameters particular to the current internal state and sensor values. It also manages the inter-object communication between the movement module and the rest of the code.

3. The highest level occurs in the behavior module itself (Chapter 12), where the decisions to initiate or continue entire type of movements are made.

5.2.1   Movement Module

The lowest level "movement module" is very robot-specific. The reader not specifically interested in the AIBO robot can safely skip to Section 5.2.2. However, for the sake of completeness, and because every robot must have *some* analogous module, we include a description of our system in this section.

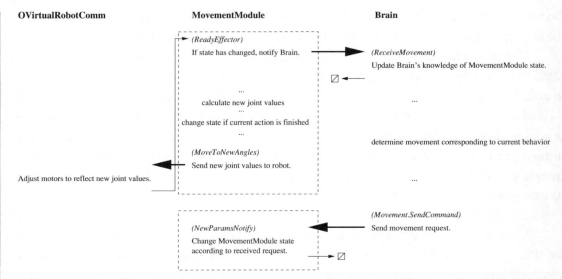

FIGURE 5.5: Inter-object communication involving the movement module. Thick arrows represent a message-containing information (from Subject to Observer); thin arrows indicate a message without further information (from Observer to Subject). An arrow ending in a null marker indicates that the message does nothing but enable the service to send another message.

The movement module shares three connections ("services") with other Open-R objects: one with the OVirtualRobotComm object mentioned above, and two with the *Brain*, the Open-R object which includes most of our code (see Chapter 10 for a description of our general architecture), including the C++ object corresponding to the movement interface described in Section 5.2.2. It uses one connection with the Brain to take requests from the Brain for types of high-level movement, such as walking in a particular direction or kicking. It then converts them to joint values, and uses its connection with OVirtualRobotComm to request that joint positions be set accordingly. These requests are sent as often as is allowed—every 8 ms. The second connection with the Brain allows the movement module to send updates to the Brain about what movement it is currently performing. Among other things, this lets the Brain know when a movement it requested has finished (such as a kick). The flow of control is illustrated by the arrows in Fig. 5.5 (the functions identified in the figure are defined below). Thick arrows represent a message-containing information (from Subject to Observer); thin arrows indicate a message without further information (from Observer to Subject). An arrow ending in a null marker indicates that the message does nothing but enable the service to send another message.

Because the movement module must send an Open-R message to OVirtualRobotComm every time it wants to change a joint position, it is necessary for the movement module to keep an

TABLE 5.1: Movement Module States

STATE	DESCRIPTION
INIT	Initial state
IDLE	No leg motion, but joint gains are set (robot is standing)[2]
PARAM_WALK	Fastest walk
SPLINE_WALK	Omnidirectional slower walk
KICK_MOTION	Kicking
GETUP_MOTION	No joint position requests being sent to OVirtualRobotComm, thus allowing built-in Sony getup routines control over all motors

internal state so that it can resume where it left off when OVirtualRobotComm returns control to the movement module. Whenever this happens, the movement module begins execution with the function ReadyEffector, which is called automatically every time OVirtualRobotComm is ready for a new command. ReadyEffector calls the particular function corresponding to the current movement module state, a variable that indicates which type of movement is currently in progress. Many movements (for example, walking and kicking) require that a sequence of sets of joint positions be carried out, so the functions responsible for these movements must be executed multiple times (for multiple messages to OVirtualRobotComm). The states of the movement module are summarized in Table 5.1.

Whereas kicking and getting up require the AIBO's head to be doing something specific, neither the idle state nor the two walks require anything in particular from the head joints. Furthermore, it is useful to allow the head to move independently from the legs whenever possible (this allows the AIBO to "keep its eye on the ball" while walking, for instance). Thus the movement module also maintains a separate internal state for the head. If the movement module's state is KICK_MOTION or GETUP_MOTION when ReadyEffector begins execution, the new joint angles for the head will be specified by the function corresponding to the movement module state. Otherwise, ReadyEffector calls a function corresponding to the current head state, which determines the new joint angles for the head, and the rest of the joint angles are determined by the function for the current movement module state. A summary of the head states appears in Table 5.2.

[2]In practice, this is implemented by executing a "walk" with forward velocity, side velocity, turn velocity, and leg height, all equal to 0.

TABLE 5.2: Head States

STATE	DESCRIPTION
IDLE	Head is still (but joint gains are set)
MOVE	Moving head to a specific position
SCAN	Moving head at a constant speed in one direction
KICK	Executing a sequence of head positions

The movement module listens for commands with a function called `NewParamsNotify`. When the Brain sends a movement request, `NewParamsNotify` accepts it and sets the movement module state and/or head state accordingly. When the internal state is next examined—this occurs in the next call to `ReadyEffector` (that is, after the next time the joint positions are set by OVirtualRobotComm)—the movement module begins executing the requested movement. See Table 5.3 for a summary of the possible requests to the movement module. Note that both a head movement and a body movement may be requested simultaneously, with the same message. However, if the body movement that is requested needs control of the head joints, the head request is ignored.

5.2.2   Movement Interface

The movement interface is part of the Brain Open-R object. Its main function is to translate high-level movement commands into movement module requests, so that the Brain can simply specify high-level movement behaviors (such as "turn toward this angle and kick with this kick") and let the movement interface take care of the rest.

During each Brain cycle, the behavior modules specify movements by calling movement interface functions, which compute the combination of movement module requests necessary to carry out the specified movement. If the requested types of movement do not interfere with each other (for example, if both a head scan and a forward walk are requested in the same Brain cycle), then all requested movements are combined in the message that is eventually sent to the movement module. Finally, at the end of each Brain cycle, the function `Movement.SendCommand` is called. This function takes care of sending the message to the movement module containing the request, and ensuring that redundant messages are not sent.

The movement interface provides functions for basic movements such as walking forward, turning, moving the head to a position, stopping the legs or head, and getting up from a

TABLE 5.3: Possible Requests to the Movement Module

TYPE OF REQUEST	EXPLANATION	ASSOCIATED PARAMETERS
MOVE_NOOP	don't change body movement	
MOVE_STOP	stop leg movement	
MOVE_PARAM_WALK	start walking using ParamWalk	x-velocity, y-velocity, angular velocity
MOVE_SPLINE_WALK	start walking using SplineWalk	x-destination, y-destination, angular destination
MOVE_KICK	execute a kick	type of kick
MOVE_GETUP	get up from a fall	
DONE_GETUP	robot is now upright, resume motions	
HEAD_NOOP	don't change head movement	
HEAD_MOVE	move head to a specific angle	
HEAD_SCAN	scan head at constant velocity	scan speed, direction
HEAD_KICK	kick with the head	type of kick
HEAD_STOP	stop head movement	

fall. It also provides several functions for more complex movements, which are described here.

Head Scan

When searching for the ball, it is helpful to move the head around in some fashion so that more of the field can be seen. On one hand, the more quickly the field can be covered by the scan, the more quickly the ball can be found. On the other hand, if the head moves too quickly, the vision will not be able to recognize the ball, because it will not be in sight for the required number of frames. Therefore, it makes sense to try to cover as much of the field with as little head movement as possible. At first, we believed that it was not possible to cover the entire height of the field with fewer than three horizontal scans, so we used a three-layer head scan at the American Open. However, by watching other teams, we became convinced that it must be possible to cover the entire relevant portion of the field with two head scans. After some experimentation, we managed to eliminate the persistent blind spot in the middle of a two-layer head scan that we created. Thus, the movement interface now provides a function that takes care of executing the two-layer head scan. It also allows the behaviors to specify which corner the scan starts from. This is because the two-layer head scan typically occurs immediately after losing the ball, and often the brain knows which direction the ball is most likely to be in given where it was last seen. Thus allowing the starting corner to be specified allows this information to be used.

Follow Object

Once the robot sees the ball, walking towards it is achieved by two simultaneous control laws. The first keeps the head pointed directly at the ball as the ball moves in the image. This is achieved by taking the horizontal and vertical distances between the location of the ball in the image and the center of the image and converting them into changes in the head pan and tilt angles.

Second, the AIBO walks towards the direction that its head is pointing. It does this by walking with a combination of forward and turning velocities. As the head's pan angle changes from the straight ahead position towards a sidewise-facing position, the forward velocity decreases linearly (from its maximum) and the turning velocity increases linearly (from zero). In combination, these policies bring the AIBO towards the ball.

While we were able to use the above methods to have the AIBO walk in the general direction of the ball, it proved quite difficult to have the AIBO reliably attain control of the ball. One problem was that the robot would knock the ball away with its legs as it approached the ball. We found that if we increased the proportionality constant of the turning velocity, it would allow the robot to face the ball more precisely as it went up to the ball. Then the ball would end up between the AIBO's front legs instead of getting knocked away by one of them. Another problem that arose was that the AIBO occasionally bumped the ball out of the way with its head. We dealt with this by having the robot keep its head pointed 10° above the ball. Both of these solutions required some experimentation and tuning of parameters.

Track Object

This function follows a ball with the head, and turns the body in place when necessary, so as not to lose sight of the ball. It is used chiefly for the goalie.

Strafe

Before we had localization in place, we needed a way to turn the robot around the ball so that it could kick it towards the goal. The problem was that we needed to keep its head pointing down the field so it could see the goal, which made turning with the ball pinched underneath the chin (see below) unfeasible. Strafing consisted of walking with a sideways velocity and a turning velocity, but no forward velocity. This caused the AIBO to walk sideways in a circle around the ball. During this time, it was able to keep its head pointed straight ahead so that it could stop when it saw the goal.

Chin Pinch Turn

This is a motion which lowers the head (to a tilt angle of $-55°$) to trap the ball below the chin, and then turns some number of degrees while the ball is trapped there. Once we had localization in place, this replaced the strafe function just described, because it is both faster and more reliable at not losing the ball.

Tuck Ball Under

This function walks forward slowly while pulling the head down. It helps the AIBO attain control of the ball, and is typically used for the transition between follow object and chin pinch turn.

5.2.3 High-Level Control

For the most part, it is the task of the behaviors to simply choose which combinations of the movement interface functions just described should be executed. However, there are exceptions; sometimes there is a reason to handle some details of movement at the level of the behavior. One such exception is establishing the duration of the chin pinch turn. Because localization is used to determine when to stop the chin pinch turn, it makes more sense to deal with this in the behavior than in the movement interface, which does not otherwise need to get localization information.

   If the behavior chooses to do a chin pinch turn (see Section 12.1.2 for details on when this happens), it will specify an AIBO-relative angle that it wishes to turn toward as well as which way to turn (by the sign of the angle). This angle is then converted to an angle relative to the robot's heading to the offensive goal.[3] The robot continues to turn[4] until the robot's heading to the opponent goal is as desired, and then the behavior transitions to the kicking state.

   While we use PARAM_WALK for the vast majority of our walking, we used SPLINE_WALK in most cases where we need to walk with a non-zero sideways velocity.

---

[3]The choice of heading to the offensive goal as the landmark for determining when the chin pinch turn should stop is due to the fact that the chin pinch turn's destination is often facing the opponent goal, as well as the fact that there was already a convenient GlobalMap interface function that provided heading to the offensive goal. In theory, anything else would work equally well.

[4]That is, the behavior repeatedly sends requests to the movement interface to execute the chin pinch turn.

An important example of this is in the supporter role (Section 13.2.1), where we need to walk to a point while facing a certain direction. SPLINE_WALK was also used for part of the obstacle avoidance challenge task. In general, we decided which walk to use in any particular situation by trying both and seeing which one was more effective.

5.3   LEARNING MOVEMENT TASKS

One major research focus in our lab is machine learning. Learning on physical robots is particularly challenging due to the limited training opportunities that result from robots moving slowly, batteries running out, etc. Nonetheless, various machine learning techniques have proven to be useful in finding control policies for a wide variety of robots including helicopters [4, 53], biped robots [91], and AIBOs [33, 34, 40]. This section presents two of our learning-research results that have been usefully incorporated into our team development during the second year.

5.3.1   Forward Gait

Our specification of the AIBO's gait left us with many free parameters that required tuning (see Appendix A.9.1). While it was possible to hand-tune these parameters as described above, we thought that machine learning could provide a more thorough and methodical approach. As shown in Fig. 5.6, our training environment consisted of multiple AIBOs walking back and

FIGURE 5.6: The training environment for the gait learning experiments. Each AIBO times itself as it moves back and forth between a pair of landmarks (A and A', B and B', or C and C').

forth between pairs of fixed landmarks. The AIBOs evaluated a set of gait parameters that they received from a central computer by timing themselves as they walked. As the AIBOs explored the gait policy space, they discovered increasingly fast gaits. This approach of automated gait-tuning circumvented the need for us to tune the gait parameters by hand, and proved to be very effective in generating fast gaits. Full details about this approach are given in [44] and video is available on-line at `www.cs.utexas.edu/~AustinVilla/?p=research/learned_walk`.

Since our original report on this method [44], there has been a spate of research on efficient learning algorithms for quadrupedal locomotion [10, 11, 13, 16, 40, 59–63]. A key feature of our approach and most of those that have followed is that the robots time themselves walking across a known, fixed distance, thus eliminating the need for any human supervision.

5.3.2   Ball Acquisition

To transition from approaching the ball to performing the chin pinch turn with it, the AIBO must *acquire* the ball so that it is securely beneath its head. This process of acquisition is very delicate and brittle, and previously has relied on repeated hand-tuning. For this reason, we chose to automate this tuning process as well, using some of the machine learning algorithms that worked well for tuning the forward gait.

The training environment for this task consists of a single AIBO on a field with a single ball. It repeatedly walks up to the ball, attempts to acquire it, and knocks it away again before starting the next trial. This was an effective way to generate a reliable acquisition behavior, and it was used with some success at competitions to accomplish the re-tuning necessitated by walk variation on the new field surfaces. Full details about our approach to learning ball acquisition can be found in [23] and video is available online at `www.cs.utexas.edu/~AustinVilla/?p=research/learned_acquisition`.

CHAPTER 6

Fall Detection

Sony provides routines that enable the robot to detect when it's fallen and that enable it to get up. Our initial approach was to simply use these routines. However, as our walk evolved, the angle of the AIBO's trunk while walking became steeper. This, combined with variations between robots, caused several of our robots to think they were falling over every few steps and to try repeatedly to get up. To remedy this, we implemented a simple fall-detection system of our own.

The fall-detection system functions by noting the robot's x- and y-accelerometer sensor values each Brain cycle. If the absolute value of an accelerometer reading indicates that the robot is not in an upright position for a number (we use 5) of consecutive cycles, a fall is registered.

It is also possible to turn fall-detection off for some period of time. Many of our kicks require the AIBO to pass through a state which would normally register as a fall, so fall-detection is disabled while the AIBO is kicking. If the AIBO falls during a kick, the fall-detection system registers the fall when the kick is finished, and the AIBO then gets up.

CHAPTER 7

Kicking

The vision and movement are generally applicable robot tasks. Any vision-based robot will need to be equipped with some of the capabilities described in Chapter 4 and any legged robot will need to be equipped with some of the capabilities described in Chapter 5. Because of the particular challenge task we adopted—robot soccer—creating a ball-kicking motion was also important. Though kicking itself may not be of general interest, it represents tasks that require precise fine motor control. In this chapter, we give a brief overview of our kick-generation process. Details of most of the actual kicks are presented in Appendix B.

The robot's kick is specified by a sequence of poses. A $Pose = (j_1, \ldots, j_n)$, $j_i \in \Re$, where j represents the positions of the n joints of the robot. The robot uses its PID mechanism to move joints 1 through n from one $Pose$ to another over a time interval t. We specify each kick as a series of moves $\{Move_1, \ldots, Move_m\}$ where a $Move = (Pose_i, Pose_f, \Delta t)$ and $Move_{j Pose_f} = Move_{(j+1) Pose_i}$, $\forall j \in [1, m-1]$. All of our kicks only used 16 of the robot's joints (leg, head, and mouth). Table 7.1 depicts the used joints and joint descriptions.

In the beginning stages of our team development, our main focus was on creating modules (Movement, Vision, Localization, etc.) and incorporating them with one another. Development of kicks did not become a high priority until after the other modules had been incorporated. However, once it became a focus issue, and after generating a single initial kick [78], we soon realized that we would need to create several different kicks for different purposes. To that end, we started thinking of the kick-generation process in general terms. This section formalizes that process.

The kick is an example of a fine-motor control motion where small errors matter. Creation of a kick requires special attention to each $Pose$. A few angles' difference could affect whether the robot makes contact with the ball. Even a small difference in Δt in a $Move$ could affect the success of a kick. To make matters more complicated, our team needed the kick to transition from and to a walk. More consideration had to be taken to ensure that neither the walk nor the kick disrupted the operation of the other.

We devised a two-step technique for kick generation:

1. Creating the kick in isolation from the walk.

2. Integrating the kick into the walk.

TABLE 7.1: Joints Used in Kicks	
JOINT	**JOINT DESCRIPTION**
j_1	front-right rotator
j_2	front-right abductor
j_3	front-right knee
j_4	front-left rotator
j_5	front-left abductor
j_6	front-left knee
j_7	back-right rotator
j_8	back-right abductor
j_9	back-right knee
j_{10}	back-left rotator
j_{11}	back-left abductor
j_{12}	back-left knee
j_{13}	head-tilt joint
j_{14}	head-pan joint
j_{15}	head-roll joint
j_{16}	mouth joint

7.1 CREATING THE CRITICAL ACTION

We first created the kick in isolation from the walk. The *Moves* that comprise the kick in isolation constitute the *critical action* of the kick. To obtain the joint angle values for each *Pose*, we used a tool that captured all the joint angle values of the robot after physically positioning the robot in its desired pose. We first positioned the robot in the *Pose* in which the robot contacts the ball for the kick and recorded $j_1, \ldots, j_n$ for that *Pose*. We called this $Pose_b$.

We then physically positioned the robot in the *Pose* from which we wanted the robot to move to $Pose_b$. We called this $Pose_a$. We then created a *Move* $m = (Pose_a, Pose_b, \Delta t)$ and watched the robot execute m. At this point of kick creation, we were primarily concerned with the path the robot took from $Pose_a$ to $Pose_b$. Thus, we abstracted away the Δt of the *Move* by selecting a large Δt that enabled us to watch the path from $Pose_a$ to $Pose_b$. We typically selected Δt to be 64. Since movement module requests are sent every 8 ms, this *Move* took 648 ms to execute.

If the *Move* did not travel a path that allowed the robot to kick the ball successfully, we then added an intermediary $Pose_x$ between $Pose_a$ and $Pose_b$ and created a sequence of two *Moves* $\{(Pose_a, Pose_x, \Delta t_i), (Pose_x, Pose_b, \Delta t_{i+1})\}$ and watched the execution. Again, we abstracted away Δt_i and Δt_{i+1}, typically selecting 64. After watching the path for this sequence of *Moves*, we repeated the process if necessary.

After we were finally satisfied with the sequence of *Moves* in the *critical action*, we tuned the Δt for each *Move*. Our goal was to execute each *Move* of the *critical action* as quickly as possible. Thus, we reduced Δt for each *Move* individually, stopping if the next decrement disrupted the kick.

7.2   INTEGRATING THE CRITICAL ACTION INTO THE WALK

The second step in creating the finely controlled action involves integrating the *critical action* into the walk. There are two points of integration: (1) the transition from the walk to the *critical action* (2) the transition from the *critical action* to the walk.

We first focus on the *Move* $i = (Pose_y, Pose_a, \Delta t)$, where $Pose_y \in \{$*all possible poses of the walk*$\}$. Since i precedes the *critical action*, there may be cases in which i adds unwanted momentum to the *critical action* and disrupts it. If i had such cases, we found a $Pose_s$, in which $\{(Pose_y, Pose_s, \Delta t), (Pose_s, Pose_a, \Delta t)\}$ did not lend unwanted momentum to the *critical action*. We call this the *initial action*. The $Pose_s$ we used mirrored the *idle* position of the walk. The *idle* position of the walk is the *Pose* the robot assumes when walking with 0 velocity. We then added the *Move* $(Pose_s, Pose_a, \Delta t)$, abstracting away the Δt, to the moves of the *critical action* and watched the path of execution.

As with the creation of the *critical action*, we then added intermediary *Poses* until we were satisfied with the sequence of *Moves* from $Pose_y$ to $Pose_a$. We then fine-tuned the Δt for the added *Moves*.

Finally, at the end of every kick during game play, the robot assumes the *idle* position of the walk, which we call $Pose_z$, before continuing the walk. This transition to $Pose_z$ takes 1 movement cycle. Thus, we consider the last *Move* of the kick, f, to be $(Pose_b, Pose_z, 1)$. Since f follows the *critical action*, there may be cases in which f hinders the robot's ability to resume walking.

In such cases, as with the creation of the *critical action* and the *initial action*, we then added intermediary *Poses* until we were satisfied with the sequence of *Moves* from $Pose_b$ to

$Pose_z$. We call the *Moves* between the intermediary *Poses* the *final action*. We then fine-tuned the values of Δt used in the *final action*.

The sequence of *Moves* constituting the *initial action*, *critical action*, and *final action* make up the kick.

Kicking has been an area of continual refinement for our team. Six of the kicks developed in the first year are summarized in Appendix B.

CHAPTER 8

Localization

One of the most fundamental tasks for a mobile robot is the ability to determine its location in its environment from sensory input. A significant amount of work has been done on this so-called *localization* problem. Since it requires at least vision and preferably locomotion to already be in place, localization was a relatively late emphasis in our efforts. In fact, it did not truly come into place until after the American Open Competition at the end of April 2003 (four months into our development effort).[1]

One common approach to the localization problem is *particle filtering* or Monte Carlo Localization (MCL) [86, 87]. This approach uses a collection of particles to estimate the global position and orientation of the robot. These estimates are updated by visual percepts of fixed landmarks and odometry data from the robot's movement module (see Section 5.1.8). The particles are averaged to find a best guess of the robot's pose. MCL has been shown to be a robust solution for mobile robot localization, particularly in the face of collisions and large, unexpected movements (e.g. the "kidnapped robot" problem [27]). Although this method has a well-grounded theoretical foundation, and has been demonstrated to be effective in a number of real-world settings, there remain some practical challenges to deploying it in a new robotic setting.

Most previous MCL implementations have been on wheeled robots with sonar or laser sensors (e.g. [26, 27]). In comparison with our setting, these previous settings have the advantages of relatively accurate odometry models and $180°$, or sometimes up to $360°$, sensory information. Although MCL has been applied to legged, vision-based robots by past RoboCup teams [45, 46, 64], our work contributes novel enhancements that make its implementation more practical.

In our first year of development, we began with a baseline implementation of Monte Carlo Localization adapted from recent literature [64] that achieves a reasonable level of competence. In the following year, we then developed a series of innovations and adjustments required to improve the robot's performance with respect to the following three desiderata:

---

[1]This chapter is adapted from [70].

1. When navigating to a point, the robot should be able to stabilize quickly close to the target destination.

2. The robot should be able to remain localized even when colliding with other objects in its environment.

3. The robot should adjust quickly and robustly to sudden large movements (the kidnapped robot problem).

All of these properties must be achieved within the limits of the robot's on-board processing capabilities.

In order to achieve these desiderata, we enhance our baseline implementation with the following three additions:

1. Maintaining a history of landmark sightings to produce more triangulation estimates.

2. Using an empirically-computed unbiased landmark distance model in addition to heading for estimate updates.

3. Tuning and extending the motion model for improved odometry calculation in a way that is particularly suited to improving localization.

We empirically evaluate the effectiveness of these general enhancements individually and collectively, both in simulation and on a Sony AIBO ERS-7 robot. In combination, the methods we present improve the robot's localization ability over the baseline method by 50%: the robot's average error in its location and heading estimates are reduced to half of that with the baseline implementation. The accuracy improvement is shown to be even more dramatic when the robot is subjected to large unmodeled movements.

8.1   BACKGROUND

In Monte Carlo Localization, a robot estimates its position using a set of samples called *particles*. Each particle, $\langle \langle x, y, \theta \rangle, p \rangle$, represents a hypothesis about the robot's *pose*: its global location (x, y) and orientation (θ). The probability, p, expresses the robot's confidence in this hypothesis. The density of particle probabilities represents a probability distribution over the space of possible poses.

Each operating cycle, the robot updates its pose estimate by incorporating information from its action commands and sensors. Two different probabilistic models must be supplied to perform these updates. The Motion Model attempts to capture the robot's kinematics. Given the robot's previous pose, and an action command, such as "walk forward at 300 mm/s" or "turn at 0.6 rad/s," the model predicts the robot's new pose. Formally, it defines the probability distribution, $p(h'|h, a)$, where h is the old pose estimate, a is the last action command executed, and h' is the new pose estimate.

The Sensor Model is a model of the robot's perceptors and environment. It predicts what observations will be made by the robot's sensors, given its current pose. The probability distribution that it defines is $p(o|h)$, where o is an observation such as "landmark X is 1200 mm away and 20 degrees to my right", and h is again the old pose estimate.

Given these two models, we seek to compute $p(h_T|o_T, a_{T-1}, o_{T-1}, a_{T-2}, \ldots, a_0)$, where T is the current cycle and h_t, o_t, and a_t are the pose estimate, observation, and action command, respectively, for time t.

8.1.1 Basic Monte Carlo Localization

The basic MCL algorithm proceeds as follows. At the beginning of each cycle, the motion model is used to update the position of each of the m particles, $\langle h^{(i)}, p^{(i)} \rangle$, based on the current action command. The new pose, $h_T^{(i)}$, is sampled from the distribution

$$h_T \sim p\big(h_T | h_{T-1}^{(i)}, a_{T-1}\big). \tag{8.1}$$

Next, the probability of each particle is updated by the sensor model, based on the current sensory data. The sensor model computes the likelihood of the robot's observations given the particle's pose, and adjusts the particle's probability accordingly. To prevent occasional bad sensory data from having too drastic an effect, a particle's change in probability is typically limited by some filtering function, $F(p_{old}, p_{desired})$. The sensor model update is given by the following equation:

$$p_T^{(i)} := F\big(p_{T-1}^{(i)}, p\big(o_T | h_{T-1}^{(i)}\big)\big). \tag{8.2}$$

Finally, particles are resampled in proportion to their probabilities. High-probability particles are duplicated and replaced with low-probability particles. The expected number of resulting copies of particle $\langle h^{(i)}, p^{(i)} \rangle$ are

$$m \cdot \frac{p^{(i)}}{\sum_{j=1}^{m} p^{(j)}}. \tag{8.3}$$

This description of the basic MCL algorithm specifies how we maintain a probabilistic model of the robot's location over time, but it omits several details. For instance, how do we obtain the motion and sensor models? And how many particles do we need? Some previous answers to these questions are surveyed in the following section.

8.1.2   MCL for Vision-Based Legged Robots

A large body of research has been performed on robot localization, mostly using wheeled robots with laser and sonar as the sensors [19, 26, 27]. Here, we focus on the few examples of MCL implemented on vision-based legged robots. In particular, our approach to localization is built upon previous research done in the RoboCup legged soccer domain [57].

Our baseline approach is drawn mainly from one particular system designed for this domain [64]. In this approach, the sensor model updates for each particle are performed based on the sensed locations of landmarks with known locations in the environment (landmarks include visual markers and line intersections in the problem domain—see Fig. 8.1). Given the particle's pose, the robot calculates the expected bearing for each observed landmark, $\alpha_{exp}^{(l)}$, $l \in L$, where L is the set of landmarks seen in the current frame. The posterior probability for a single observation is then estimated by the following equation representing the degree to which the observed landmark bearing $\alpha_{meas}^{(l)}$ matches $\alpha_{exp}^{(l)}$:

$$s\left(\alpha_{meas}^{(l)}, \alpha_{exp}^{(l)}\right) = \begin{cases} e^{-50\omega_l^2}, & \text{if } \omega_l < 1 \\ e^{-50(2-\omega_l)^2}, & \text{otherwise} \end{cases} \tag{8.4}$$

where $\omega_l = \frac{\left|\alpha^{(l)}{}_{(meas}-\alpha_{exp}^{(l)})\right|}{\pi}$. The probability, p, of a particle is then the product of these similarities:

$$p = \prod_{l \in L} s\left(\alpha_{meas}^{(l)}, \alpha_{exp}^{(l)}\right). \tag{8.5}$$

Lastly, the following filtering function is applied to update the particle's probability [64]:

$$p_{new} = \begin{cases} p_{old} + 0.1, & \text{if } p > p_{old} + 0.1 \\ p_{old} - 0.05, & \text{if } p < p_{old} - 0.05 \\ p, & \text{otherwise.} \end{cases} \tag{8.6}$$

An important aspect of this sensor model is that distances to landmarks are completely ignored. The motivation for this restriction is that vision-based distance estimates are typically quite noisy and the distribution is not easy to model analytically. Worse yet, there can be a strong nonlinear bias in the distance estimation process which makes the inclusion of distance estimates actively harmful, causing localization accuracy to degrade (see Section 8.3). In this section, it is shown that the bias in distance estimates can be empirically modeled such that they can be incorporated to improve sensor model updates.

In the baseline approach, to address the frequently-occurring kidnapped robot problem,[2] a few of the particles with low probability are replaced by estimates obtained by triangulation from the landmarks seen in the current frame. This process, called *reseeding*, is based upon the idea of Sensor Resetting localization [46].

A shortcoming of previous reseeding approaches is that they require at least two or three landmarks to be seen in the same camera frame to enable triangulation. This section presents a concrete mechanism that enables us to use reseeding even when two landmarks are never seen concurrently.

Another significant challenge to achieving MCL on legged robots is that of obtaining a proper motion model. In our initial implementation, we used a motion model that provided reasonably accurate velocity estimates when the robot was walking at near-maximum speed. However, when the robot was close to its desired location, moving at full speed, caused jerky motion. The resulting "noise" in the motion model caused erroneous action model updates (Eq. (8.1)). The robot assumed that it was moving at full speed, but before its motion was completed, it received a pose estimate beyond the target. This estimate generated another motion command, leading to oscillation around the target position. This section examines the effect on this oscillation of improving the motion model with an extension to the robot's behavior.[3]

8.2   ENHANCEMENTS TO THE BASIC APPROACH

This section details the three enhancements that we made to our baseline implementation to obtain significant improvements in the robot's localization accuracy. These enhancements are each individually straightforward, and they do not change the basic particle filtering approach. But together they provide a roadmap for avoiding potential pitfalls when implementing it on the vision-based and/or legged robots.

8.2.1   Landmark Histories

To triangulate one's pose from the fixed landmarks, either two or three landmarks must be seen, depending on whether or not distance information is used. It is possible to reseed without seeing enough landmarks simultaneously by maintaining a *history* of previous observations. Observed distances and angles to landmarks are stored over successive frames. The stored distances and angles are adjusted each frame based on the robot's known motion. Successive

---

[2]In the RoboCup legged league, when a robot commits a foul, it is picked up by the referee and replaced at a different point on the field.

[3]Note that the noisy motion model is not a property of MCL or any other algorithm, but rather our own baseline implementation.

observations of the same landmark are averaged, weighted by their confidence, then given as input for reseeding, as described in Section 8.1.

More precisely, for each of the N landmarks, let M_i be the number of times the robot has recently observed landmark i. We represent the jth observation of the ith landmark as $Obs_{i,j} = (d_{i,j}, o_{i,j}, p_{i,j}, t_{i,j})$, where d and o are the relative distance and orientation of the landmark, t is the timestamp of this observation, and p is the probability of the observation according to a vision-based confidence measure [79]. Also, let $\overrightarrow{pos_{i,j}}$ be the two-dimensional Cartesian vector representation of the observation relative to the robot.

Given a 2-D velocity vector representing the robot's current motion, $\vec{v}$, the change in position of the robot is given by

$$\overrightarrow{\delta pos} = \vec{v} * (t_c - t_{lu}) \tag{8.7}$$

where t_c and t_{lu} represent the current time and the time of the last update. Then, observations are *corrected* as

$$\overrightarrow{pos_{i,j}}\big|_{\substack{i\in[1,N] \\ j\in[1,M_i]}} = \overrightarrow{pos_{i,j}} - \overrightarrow{\delta pos}. \tag{8.8}$$

Next, to merge the observations corresponding to any one landmark i, we obtain the distance, heading, and probability of the aggregate landmark observations as

$$psum_i = \sum_j p_{i,j}, \qquad p_i = \frac{psum_i}{M_i}$$
$$d_i = \frac{\sum_j p_{i,j}d_{i,j}}{psum_i}, \qquad o_i = \sum_j p_{i,j}o_{i,j}. \tag{8.9}$$

Because the motion model is not very accurate and the robot can be picked up and moved to a different location frequently, the history estimates are deleted if they are *older* than a threshold in time or if the robot has undergone significant movement (linear and/or rotational). That is, we remove observations from the history if any one of the following three conditions are true:

$$t_{i,j} \geq t_{th}, \qquad d_{i,j} \geq d_{th}, \qquad o_{i,j} \geq o_{th} \tag{8.10}$$

In our current implementation, we use $t_{th} = 3$ s, $d_{th} = 15.0$ cm, and $o_{th} = 10.0°$. These thresholds were found through limited experimentation, though the algorithm is not particularly sensitive to their exact values.

8.2.2 Distance-Based Updates

To use distances to landmarks effectively in localization, we must first account for the nonlinear bias in their estimation. Initially, the estimation was performed as follows:

- The landmarks in the visual frame are used to arrive at displacements (in pixels) with respect to the image center.

- Using similar triangles, along with knowledge of the camera parameters and the actual height of the landmark, these displacements are transformed into distance and angle measurements relative to the robot.

- Finally, we transform these measurements, using the measured robot and camera (tilt, pan, roll) parameters to a frame of reference centered on the robot.

Using this analytic approach, we found that the distances were consistently underestimated. The bias was not constant with distance to the landmarks, and as the distance increased, the error increased to as much as 20%. This error actually made distance estimates harmful to localization.

To overcome this problem, we introduced an intermediate correction function. We collected data corresponding to the measured (by the robot) and *actual* (using a tape measure) distances to landmarks at different positions on the field. Using polynomial regression, we estimated the coefficients of a cubic function that when given a measured estimate, provided a corresponding *corrected* estimate. That is, given measured values X and actual values Y, we estimated the coefficients, a_i, of a polynomial of the form

$$y_i|_{y_i \in Y} = a_0 + a_1 x_i + a_2 x_i^2 + a_3 x_i^3 \cdot |_{x_i \in X}. \qquad (8.11)$$

During normal operation, this polynomial was used to compute the *corrected* distance to landmarks. Once this correction was applied, the distance estimates proved to be much more reliable, with a maximum error of 5%. This increased accuracy allowed us to include distance estimates for both probability updates and reseeding.

8.2.3 Extended Motion Model

In our baseline implementation, an inaccurate motion model prevented the robot from being able to precisely navigate to a specific location on the field, a desirable goal. To overcome this obstacle, we modified the robot's behavior during localization. When navigating to a point, the robot moves at full speed when it is more than a threshold distance away. When it comes closer to the target position, it progressively slows down to a velocity almost $\frac{1}{10}$ the normal speed. The distance threshold was chosen to be 300 mm based on experimentation to find a good trade-off between accuracy and stabilization time. The robot's performance is not very sensitive to this value.

Although this is a minor contribution to our overall set of enhancements, a properly calibrated motion model can lead to a considerable decrease in oscillation, which significantly improves the localization accuracy. Also, as shown below, reduced oscillation leads to increased accuracy and smoother motion while not increasing the time to stabilize.

8.3   EXPERIMENTAL SETUP AND RESULTS

This section describes our experimental platform and the individual experiments we ran to measure the effect of these enhancements on localization performance.

8.3.1   Simulator

Debugging code and tuning parameters are often cumbersome tasks to perform on a physical robot. Particle filtering implementations require many parameters to be tuned. In addition, the robustness of the algorithm often masks bugs, making them difficult to track down. To assist us in the development of our localization code, we constructed a simulator. Conveniently, the simulator has also proven to be useful for running experiments like the ones presented below.

The simulator does not attempt to simulate the camera input and body physics of the actual Sony AIBO. Instead, it interacts directly with the localization level of abstraction. Observations are presented as distances and angles to landmarks relative to the robot. In return, the simulator expects high-level action commands to move the robot's head and body. The movement and sensor models both add Gaussian noise to reflect real-world conditions. Details can be found in Chapter 14.

8.3.2   Experimental Methodology

According to the desiderata presented in Chapter 8, we set out to evaluate the robot's localization performance based on

- Overall accuracy;
- Time taken to navigate to a series of points;
- Ability to stabilize at a target point; and
- Ability to recover from collisions and "kidnappings."

We devised a group of experiments to measure the effects of our enhancements with regard to these metrics. Though we ran as many experiments as possible on the actual robot,

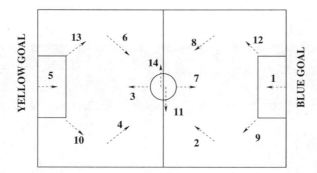

FIGURE 8.1: Points, numbered in sequence, that the robot walks during experimentation; arrows indicate the robot's target heading.

we found it necessary to use the simulator for the recovery experiments because it allowed us to consistently reproduce collisions and kidnappings while accurately tracking the robot's true pose over time.

8.3.3 Test for Accuracy and Time

To test the effect of the incorporated enhancements on overall accuracy and time, we designed a task in which the robot was required to visit a sequence of 14 points on the field as depicted in Fig. 8.1. The robot was allowed to repeatedly scan the environment.[4] For each run, we measured the time taken and the error in position and angle at each point.

To measure the individual effects of the added enhancements, we performed this task over six different conditions:

1. Baseline implementation (none).
2. Baseline + Landmark Histories (HST).
3. Baseline + Distance-based probability updates (DST).
4. Baseline + Function approximation of distances (FA).
5. Baseline + Function approximation of distances and distance-based probability updates (DST + FA).
6. Baseline + All enhancements (All).

These conditions were chosen to test each enhancement in isolation and in combination. Further, it allows us to demonstrate that using landmark distances can be harmful if the distance estimates are not properly corrected. In all the six conditions, we used the

---

[4]In general, the robot is not able to scan the field constantly—we allow it to do so in these experiments for the purpose of consistency.

TABLE 8.1: Errors in Position and Orientation

ENHAN.	DISTANCES		ANGLE	
	ERROR (CM)	P-VALUE	ERROR (DEG)	P-VALUE
None	19.75 ± 12.0	–	17.75 ± 11.48	–
HST	17.92 ± 9.88	0.16	10.68 ± 5.97	10^{-10}
DST	25.07 ± 13.73	10^{-4}	9.14 ± 5.46	10^{-13}
FA	15.19 ± 8.59	10^{-4}	10.21 ± 6.11	10^{-11}
DST + FA	13.72 ± 8.07	10^{-6}	9.5 ± 5.27	10^{-13}
All	9.65 ± 7.69	10^{-15}	3.43 ± 4.49	$<10^{-15}$

behavior-based motion model enhancement (Section 8.2.3). The impact of the extended motion model was tested separately on a task requiring especially accurate positioning (see Section 8.3.5).

The results of these experiments, averaged across ten runs each, are shown in Tables 8.1 and 8.2. The localization errors were computed as the distance between the robot's center and the target location when the robot indicated that it believed it had reached the target. Significance is established using a *Student's t-test*. The p-values measure the likelihood that

TABLE 8.2: Average Time Taken Per Run

ENHAN.	TIME (S)	P-VALUE
None	161.25.75 ± 3.43	–
HST	161.26 ± 5.96	0.75
DST	196.18 ± 12.18	10^{-6}
FA	171.85 ± 15.19	0.04
DST + FA	151.28 ± 48.06	0.56
All	162.54 ± 4.38	0.43

each entry differs from the baseline algorithm (labeled *None*). We follow the convention of using a p-value of < 0.05 (>95% confidence) to establish statistical significance.

From Table 8.1, it is evident that the error in pose is considerably reduced after the addition of all the enhancements. There is a 50% reduction in position error with respect to the baseline implementation. The improvement in orientation accuracy is even more dramatic.

Though the addition of observation histories alone (*HST*) does not significantly improve accuracy, when combined with the other enhancements, there is a significant improvement (*p-value* between *DST* + *FA* and *All* $\leq 10^{-15}$).

The general reluctance to utilize distance-based probability updates in localization is explained by the fact that when they are used without accounting for the nonlinear bias through function approximation, the error is more than even the baseline implementation. By itself, *FA* produces a good reduction in error because the improved distance estimates lead to better reseed values, even if they are not used for probability updates. Using both, i.e. *DST* + *FA*, results in the biggest improvement (*p-value* $= 10^{-15}$ w.r.t. DST) after *All*.

From Table 8.2, we see that the addition of all the enhancements does not significantly increase the time taken to reach the target locations. However, using *DST* without incorporating *FA* or *HST* does lead to worse time performance because the robot has trouble settling at a point. Since the oscillation happens even with the extended motion model in place, it can be attributed solely to bad probability updates. Though the average time taken is the lowest for *DST* + *FA*, this was not significant (p-value between *DST* + *FA* and *All* is 0.51).

8.3.4   Test for Stability

In addition to being able to localize precisely, once the robot has arrived at the target location, it should stay localized, a property we refer to as *stability*. To test stability, we developed the following experiment. The robot is placed at each one of the points shown in Fig. 8.1 and is given 10 s to localize. Subsequently, for 20 s, the robot remains stationary and records its pose estimates, at the end of which period the robot calculates the deviation in its pose estimates. Since the robot does not actually move during this period, changing pose estimates reflect erroneous oscillation.

Table 8.3 summarizes the results. The values shown in the table are the average deviation in pose estimates over ten runs at each point. These 140 data points reflect the average deviation obtained from roughly 600 samples each.

Based on these results, we can conclude that the addition of all enhancements provides a significant improvement in stability. The use of *DST* without the other two enhancements (*HST*, *FA*) once again performs the worst. It is surprising that *FA* does as well as *All* ($p_{dist} = 0.413$, $p_{ang} = 0.743$). Then again, because the distances are being estimated well, the robot

TABLE 8.3: Average Deviation in Position and Orientation

ENHAN.	DIST ERROR (CM)	P-VALUE	ANG ERROR (DEG)	P-VALUE
None	2.63	–	0.678	–
HST	1.97	10^{-5}	0.345	$<10^{-15}$
DST	9.26	$<10^{-15}$	3.05	$<10^{-15}$
FA	1.46	10^{-10}	0.338	$<10^{-15}$
DST + FA	4.07	10^{-8}	1.30	$<10^{-15}$
All	1.32	10^{-12}	0.332	$<10^{-15}$

gets better reseed estimates. This confirms our hypothesis that landmark distances can be useful enhancements.

8.3.5   Extended Motion Model

We performed an additional experiment to determine the importance of a well-calibrated and extended motion model. In the robot soccer domain, one of the robots typically plays the role of keeper and must stay well-localized within the goal box (points 1 and 5 in Fig. 8.1). Doing so requires precise and controlled movements. We performed an experiment in which the robot, playing the role of a keeper, tries to stay localized at its base position. We repeatedly moved the robot out of its position to the center of the field and let it move back to its base position. We then measured the error in the pose estimate over ten runs of this test both with and without an extended motion model. The results are shown in Table 8.4.

From the results in Table 8.4, we see that incorporating an extended motion model significantly boosts the localization performance. The corresponding increase in time is not large enough to be disruptive.

TABLE 8.4: Errors With and Without the Extended Motion Model

ENHAN.	ERROR		
	DIST (CM)	ANG (DEG)	TIME (S)
None	12.89 ± 2.48	15.0 ± 9.72	17.21 ± 1.25
Extended MM	7.496 ± 1.96	5.5 ± 4.97	18.14 ± 2.25

8.3.6 Recovery

Finally, we performed a set of experiments using our simulator to test the effect of our enhancements on the robot's ability to recover from unmodeled movements. We tested our localization algorithm against two types of interference that are commonly encountered in RoboCup soccer: collisions and kidnappings. In both cases, we disrupt the robot once every 30 s (simulated) while it attempts to walk a figure-8 path around the field and scan with its head. Collisions are simulated by preventing the robot's body from moving (unbeknownst to it so it continues to perform motion updates) for 5 s. The robot's head is permitted to continue moving. In the kidnapping experiments, the robot is instantly transported to a random spot on the field, 1200 mm from its previous location and given a random orientation. We test our enhancements by comparing these scenarios to the case in which the robot simply walks a figure-8 pattern undisturbed. Twice per second, we record the absolute error in robot's pose estimate. We compare average errors for 2 h of simulated time, corresponding to roughly 50 laps around the field. The results are summarized in Tables 8.5 and 8.6.

From the distance error table, we can see that for every test condition, the collision and kidnapping disturbances caused an increase in average error when compared to the undisturbed scenario, as we would expect. However, the increase in error is much smaller when our enhancements were used. For instance, with all enhancements turned off, the kidnapped robot problem causes almost a 10-fold increase in error. But when we include our enhancements, the error is only increased by 56%. There is a corresponding improvement in angle accuracy.

When we look at the individual contributions of the enhancements, we see that for both disturbance scenarios, the individual enhancements are better than no enhancements. Also, in both cases, they have a larger effect when combined than when they are used individually. This implies that both enhancements contribute to the improved recovery. Only in the case

TABLE 8.5: Errors for Various Perturbations

	DISTANCE ERROR (CM)		
ENHAN.	UNDISTURBED	COLLIDING	KIDNAPPED
None	8.03 ± 4.92	27.7 ± 22.4	74.3 ± 55.2
HST	17.6 ± 16.2	25.3 ± 21.5	27.3 ± 36.4
DST + FA	7.83 ± 5.35	16.2 ± 16.9	31.5 ± 41.6
All	8.67 ± 9.68	14.4 ± 15.6	13.5 ± 23.4

TABLE 8.6: Errors for Various Perturbations

	ANGLE ERROR (DEG)		
ENHAN.	UNDISTURBED	COLLIDING	KIDNAPPED
None	2.74 ± 2.20	7.15 ± 9.33	15.3 ± 22.7
HST	3.69 ± 3.15	10.7 ± 21.4	6.66 ± 15.3
DST + FA	2.91 ± 2.36	7.07 ± 11.0	9.81 ± 21.2
All	2.38 ± 2.31	5.57 ± 11.1	4.38 ± 13.5

of *DST + FA*, when comparing angle accuracy to the baseline for collisions, do we not see a significant improvement. This is most likely explained by the limited effect of distance updates on orientation accuracy in general.

Finally, looking at the "Undisturbed" column, we notice that the enhancements, if anything, have a somewhat negative effect in the absence of interference. It would seem that this finding is incompatible with the accuracy results reported in Table 8.1. However, the experimental conditions in the two scenarios were significantly different. In particular, in the recovery experiments, we did not wait for the robot's estimate to stabilize before measuring the error. This metric not only measures how well the robot localizes when it has good information, but also how poorly it does when it is lost, which is why this metric is appropriate for testing the recovery.

Although the distance error for *HST* is quite high, in the *All* case we do only slightly worse than originally. Thus, although the algorithm has been tuned to work well in game situations where collisions and kidnappings are common, we don't give up much accuracy when those disturbances do not occur.

8.4   LOCALIZATION SUMMARY

This section addressed the task of performing accurate localization on a legged robot with vision as the primary sensor. This task presents new challenges in terms of sensor and motion modeling in comparison with previous approaches on wheeled robots with a ring of range sensors. Starting with a *baseline* implementation adapted from the literature, several novel enhancements were presented. Empirical results were presented, both on a physical robot and in simulation that demonstrate that these enhancements improve the robot's overall accuracy, its stability when stationary, and its recovery from disturbances. This work contributes a detailed cataloging of some potential pitfalls and their solutions when implementing particle filtering on legged robots with limited-range of vision.

Though this section focusses on MCL, or particle filtering, approaches to localization, which have been popular in recent years; other effective approaches also exist and have been used on AIBOs and on other robots as well. One such example is the Extended Kalman Filter (EKF), a modification of the classic Kalman Filter [38] to handle nonlinearities in the system. The EKF has been extensively used for state estimation. On robots, it has been used for pose estimation [31, 35, 47, 48] using both range data (from laser or sonar) and visual data (from a camera). It has been used for localization in a distributed setting [49], and also for the problem of simultaneous localization and mapping [37].

On the AIBOs in particular, it has been used successfully by several teams [17, 18, 66], including the NuBots who won the four-legged RoboCup Competition in 2006. Other teams have used a combination of particle filters and EKF for localization [7]. Gutmann and Fox [30] provide a good comparison of the various localization algorithms tested on data obtained from the robot soccer setting.

CHAPTER 9

Communication

Collective decision making is an essential aspect of a multiagent domain such as robot soccer. The robots thus need the ability to share information among themselves. This chapter discusses the methodologies we adopted to enable communication and the various stages of development of the resulting communication module.

9.1 INITIAL ROBOT-TO-ROBOT COMMUNICATION

Our initial goal was to understand the capabilities and limitations of the wireless communication channel between the various robots. We created a simple server and a client that used the User Datagram Protocol (UDP). We chose UDP because it typically provides greater bandwidth than the alternative, TCP. Our intent was to determine how quickly we could transfer data between robots and to simply get used to writing applications that would allow the robots to communicate.

The first server that we created generated a few bytes of data and tried to broadcast it to a client. The client program simply gathered these data as it received it. We ran the server and the client on two different robots and monitored their actions by telnetting into them.

Once that worked, we extended our communication modules to interface with the robot's mechanical parts. The next server that we created captured the joint angles of the robot and broadcast them to the client. The client gathered the data and set its own joint positions accordingly. Thus, when we moved the legs of the server robot, the client robot would move its legs by the same amount, thus acting as a master–slave (puppet) interface.

As we became familiar with the networking interfaces of the robot, we continued to explore the various uses of communication. We streamed images from the robot's camera to a PC with both UDP and TCP, created a hierarchy of single-master, multiple-slaves that enabled one robot to "lead" a team of robots, and coded a remote-control program that we could use to control the AIBO from a PC. All of these experiments provided valuable feedback that we later used when creating both our full robot-to-robot communication module (described below) and UT Assist (Chapter 15).

Ultimately, we needed to interface with an externally-provided module called TCP-Gateway, that was used in the RoboCup competitions and that abstracted away most of the low-level networking, providing a standard Open-R interface in its place. Because the details of this interface are specific to the AIBO robot, these are described only in Appendix C.

9.2   MESSAGE TYPES

One of the challenges we faced regarding communication was the possibility that multiple types of messages would need to be sent. We could theoretically handle this with a stage in the brain loop that could read and distribute messages appropriately. As we proceeded, however, this option became more and more unwieldy. Variables and data that would be used in one part of a program would be read and set in another part, perhaps even in another file. What we needed was the ability to create an arbitrary number of different message types, such that anywhere in the program, we could request from the communication system the next message *of that type*.

Our first implementation kept the same communication stack, but when a request was made, the type of message was passed as a parameter. The communication system would then search through the stack for the next message of that type, remove it from the stack, and return it. This worked fine, but we quickly realized that if any one type of message ceased to be consumed, it could have serious ramifications in terms of the time needed to retrieve other types of messages.

To solve this, we implemented an array of communication stacks, one for each type of message. This gave us a constant-time fetch for the next message of any type. As messages arrived, they were processed by their type and placed into the correct stack. This way, messages related to global maps could be retrieved and used in the code that actually handles the operation of global maps, while messages relating to strategy changes could be handled in a different part of the code.

9.3   KNOWING WHICH ROBOTS ARE COMMUNICATING

As we played more games with our robots, we noticed that sometimes there were communication errors because two robots for some reason or another were not connected to one another. Unfortunately, this kind of error would only be detected when we noticed the robots behaving strangely—something that would not show up until well into a game. Because of the large negative effect of communication failures in our game-play, we decided that we required a way to determine which robots were connected to each other, such that we would have a chance to correct it before starting the game. Additionally, if communication links were to fail during a game, we want our robots to notice this and be able to incorporate this knowledge into their behavior.

Initially, we had a system that allowed us to tell if a robot was "hearing" from any other robot—a blue LED on the top of the head would blink every time a message was received. This was not sufficient—a robot could be hearing from only one other robot, but missing out on the important information from the other 2 robots. In reality, the system is much more complex than this. With 2 connections between each robot (each direction is a separate connection), there are a total of 12 connections of which to keep track. Eventually, we decided to light up four LEDs on the face (the new ERS-7 robots made this possible), one for each robot. If the robot believes another robot to be "dead", that is, it hasn't heard from it in a while, then that LED is not illuminated. This provides us with a very quick and easy way to establish both whether communication is present (if the lights are illuminated) and what the quality of the communication is (if the lights are blinking in and out).

9.4   DETERMINING WHEN A TEAMMATE IS "DEAD"

In order to determine whether or not a robot has heard from another robot lately, each robot keeps track of how many brain cycles ago it last heard from each other robot. If this exceeds a certain quantity, which we called START_PINGING_THRESHOLD (50), the robot will then send the other robot a new message type, called a PING_MESSAGE. Whenever a robot receives a PING_MESSAGE, it is required to immediately respond with a PONG_MESSAGE. If the first robot receives the PONG_MESSAGE, it resets the counter for the last time it heard from the other robot, and things proceed normally. However, if it fails to hear back by a specified interval, CYCLES_BETWEEN_PINGS (25), another PING_MESSAGE will be sent. This will continue until a PONG_MESSAGE is received. If the count of cycles since it last heard from the other robot exceeds the threshold CONSIDER_DEAD_THRESHOLD (100), the other robot is considered "dead" by the first robot. This means that it will act as if the other robot does not exist, making decisions that reflect a 3-, 2-, or 1-robot strategy. If after this point, the first robot receives a PONG_MESSAGE (or any other message), it will restore the other robot's "alive" status. By varying these parameters, we can control how sensitive and responsive the robots are to communication failures.

9.5   PRACTICAL RESULTS

This became very useful in the World Cup competition, where communication experienced very high latencies. It allowed our robots to cope reasonably well with the communication problems as well as adjust their strategies accordingly. The main problem we experienced was that one of the other leagues was transmitting on a frequency very close to ours. This caused our communications to be delayed by sometimes up to 10 or 20 s. With our original configuration, our robots would not have known what to do—they would have been stuck waiting for instructions from other robots. However, because they could tell that their teammates were *incommunicado*, they were able to act independently.

CHAPTER 10

General Architecture

Due to our bottom-up approach, we did not address the issue of general architecture until some important modules had already taken shape. We had some code that cobbled together our vision and movement components to produce a rudimentary but functional goal-scoring behavior (see Section 12.1.1). Although this agent worked, we realized that we would need a more structured architecture to develop a more sophisticated agent, particularly with the number of programmers working concurrently on the project. The decision to adopt the architecture described below did not come easily, since we already had something that worked. Implementing a cleaner approach stopped our momentum in the short-term and required some team members to rewrite their code, but we feel the effort proved worthwhile as we continued to combine more independently-developed modules.

We designed a framework for the modules with the aim of facilitating further development. We considered taking advantage of the operating system's inherent distributed nature and giving each module its own process. However, we decided that the task did not require such a high degree of concurrent computation, so we organized our code into just two separate concurrent objects (Fig. 10.1).

We encapsulated all of the code implementing low-level movement (Section 5.2.1) in the MovementModule object. This module receives Open-R messages dictating which movement to execute. Available leg movements include locomotion in a particular direction, speed, and turning rate; any one of a repertoire of kicks; and getting up from a fallen position. Additionally, the messages may contain independent directives for the head, mouth, and tail. The MovementModule translates these commands into sequences of set points, which it feeds as messages into the robot's OVirtualRobotComm object. Note that this code inhabits its own Open-R object precisely so that it can supply a steady stream of commands to the robot asynchronously with respect to sensor processing and deliberation. For further details on the movement module, see Section 5.2.1.

The Brain object is responsible for the remainder of the agent's tasks: accepting messages containing camera images from OVirtualRobotComm, communicating over the wireless network, and deciding what movement command messages to send to the MovementModule object. It contains the remaining modules, including Vision, Fall Detection,

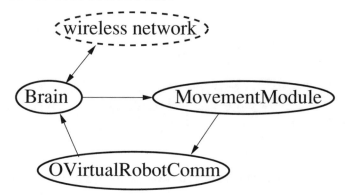

FIGURE 10.1: A high-level view of the main Open-R objects in our agent. The robot sends visual data to the Brain object, which sends movement commands to the MovementModule object, which sends set points to the PID controllers in the robot. The Brain object also has network connections to teammates' Brain objects, the RoboCup game controller, and our UT Assist client (Chapter 15). Note that this figure omits sensor readings obtained via direct Open-R API calls.

Localization, and Communication. These components thus exist as C++ objects within a single Open-R object. The Brain itself does not provide much organization for the modules that comprise it. In large part, it serves as a container for the modules, which are free to call each other's methods.

From an implementation perspective, the Brain's primary job is to activate the appropriate modules at the appropriate times. Our agent's "main loop" activates whenever the Brain receives a new visual image from OVirtualRobotComm. Other types of incoming data, mostly from the wireless network, reside in buffers until the camera instigates the next Brain cycle. Each camera image triggers the following sequence of actions from the Brain:

Get Data: The Brain first obtains the current joint positions and other sensor readings from Open-R. It stores these data in a place where modules such as Fall Detection can read them. This means that we ignore the joint positions and sensor readings that OVirtualRobotComm generates between vision frames.

Process Data: Now the Brain invokes all those modules concerning interpreting sensor input: Vision, Localization, and Fall Detection. Note that for simplicity's sake even Communication data wait until this step, synchronized by inputs from the camera, before being processed. Generally, the end-result of this step is to update the agent's internal representation of its external environment: the global map (see Chapter 11).

Act: After the Brain has taken care of sensing, it invokes those modules that implement acting, described in Chapters 12 and 13. These modules typically don't directly access the data gathered by the Brain. Instead they query the updated global map.

CHAPTER 11

Global Map

Early in the development of our soccer-playing agent, particularly before we had functioning localization and communication, we chose our actions using a simple finite state machine (see Chapter 12). Our sensory input and feedback from the Movement Module dictated state transitions, so sensations had a relatively direct influence on the behavior. However, once we developed the capability to locate our agents and the ball on the field and to communicate this information, such a direct mapping became impossible. We created the global map to satisfy the need for an intermediate level of reasoning. The global map combines the outputs of Localization from Vision and from Communication into a coherent picture of what is happening in the game, and it provides methods that interpret this map in meaningful ways to the code that governs behavior.

11.1 MAINTAINING LOCATION DATA

When a robot computes new information about the location of any particular object on the field, it usually merges the new estimate of position with the current estimate of position that is stored in its global map (see Section 4.6).

As time passes, the error estimate for all of the information in the global map increases. This degradation of information is included to more accurately model the rapid rate of change in the state of the game. The idea is to make the degradation smooth to reflect the maximum change that we are ready to allow (i.e., the change that we think could have happened) since the last update. The approach used here is to estimate a maximum 'velocity' by which we assume the object can move along the x- and y-axes. We then use this velocity to calculate the maximum distance the object could have moved along the axes in the time since the last update. The estimated change, σ_{change}, is statistically added to the location's uncertainty in accordance with the formula

$$\sigma_{\text{updated}} = \sqrt{\sigma_{\text{previous}}^2 + \sigma_{\text{change}}^2}. \tag{11.1}$$

For example, if we consider the modeling of the opponents, we want our estimates of the opponents to be as accurate as possible and we do not want new estimates to occur every frame.

We would ideally want to be able to merge estimates from the current frame with those in the previous frame, wherever possible, so that we can actually map the motion of the opponents. At the same time, we may have spurious detections every once in a while and if they are not seen in successive frames, we want these estimates to disappear quickly. So for opponents we use an artificially high 'velocity' such as 1500 mm/s (determined by experimentation). On the other hand, we want the estimates of the ball, robot position, and those of the teammates to degrade depending on some 'velocity' that reflects their actual motion. So we choose the velocity for teammate motion as 300 mm/s (we do not think any team can move any faster than that as yet) and that for the ball as 1000–1500 mm/s because the ball can move about that fast after a single powerful kick. These values were all determined experimentally and seem to provide reasonable performance in terms of how we would like our estimates to be updated.

11.2   INFORMATION FROM TEAMMATES

Each robot periodically sends information from its global map to each of its teammates. This transmitted information includes the following.

1. The location of the robot, along with an error estimate.
2. The locations of any opponents of which the robot currently is aware, along with error estimates.
3. The location of the ball, along with an error estimate.

When robot A receives teammate position information from robot B, robot A always assumes that B's estimate of B's position is better than A's estimate of B's position. Therefore, robot A simply replaces its old position for B with the new position.

When a robot receives opponent information from another robot, it updates its current estimate of opponent locations as described in Section 4.6.

If robot A has seen the ball recently when it receives a ball position update from robot B, robot A ignores B's estimate of ball position. If robot A hasn't seen the ball recently, then it merges its current estimate of the ball's position with the position estimate that it receives from robot B.

The basic idea behind having a global map is to make sharing of information possible so that the team consisting of individual agents with limited knowledge of their surroundings can pool the information to function better as a team. The aim is to have completely shared knowledge but the extent to which this succeeds is dependent upon the ability to communicate. Since the communication (see Chapter 9) is not fully reliable, we have to design a good strategy (Chapter 12 describes our strategy and behaviors) that uses the available information to the

TABLE 11.1: The Predicates That Global Map Provide

getID	GetDistanceFromDefender	InLeftThird
getTeamMembers	GetDistanceFromKeeper	InCentralThird
getOpponents	GetAttackerRelativePosition	InRightThird
getBall	GetSupporterRelativePosition	InTopQuarter
getMyPosition	GetDefenderRelativePosition	InOwnHalf
adjustRelativeBall	GetKeeperRelativePosition	IsLower
wellLocalized	GetAttackerAbsolutePosition	InOwnGoalBox
ballOnField	GetSupporterAbsolutePosition	AmILeftMost
getBallDistanceFromOurGoal	GetDefenderAbsolutePosition	AmIRightMost
getRelativeBall	GetKeeperAbsolutePosition	GetLeftPosAngle
getRelativeOrientation	IsAttackerWellLocalized	GetRightPosAngle
getRelativeOpponentGoal	IsSupporterWellLocalized	OpponentsOnLeft
getRelativeOwnGoal	IsDefenderWellLocalized	OpponentsOnRight
getRelativeOpponents	IsKeeperWellLocalized	NumOpponentsOnLeft
getRelativeTeamMembers	BallInOwnGoalBox	NumOpponentsOnRight
GetRelativePositionOf	BallInOppGoalBox	OnOurSideOfTheField
GetRelativePositionOfTeamRel	BallInOurHalf	OnLeftSideOfTheField
HeadingToOffPost	AmIInDefensiveZone	IAmClosestTo
HeadingToDefPost	NearOwnGoalBox	IAmClosestToBall
GetClosestCorner	NumberOfTeamMatesInOpponentHalf	NumOpponentsWithinDistance
DistanceToOffPost	NumberOfTeamMatesInOwnHalf	GetRelativePositionTo
DistanceToDefPost	HeadingToOppGoal	InZone
GetDefensivePost	HeadingToOwnGoal	ApproachingZone
GetDistanceFromAttacker	HeadingToOppLeftCorner	
GetDistanceFromSupporter	HeadingToOppRightCorner	

maximum extent possible. Other modules can access the information in the GlobalMaps using the accessor functions (predicates) described in the following section.

11.3   PROVIDING A HIGH-LEVEL INTERFACE

From a high-level perspective, the only data that the global map provides to other modules are the estimated positions of the ball and the robots on the field, along with degrees of uncertainty about these estimates. However, the global map also houses an array of functions on these data, to prevent different portions of the behavior code from replicating commonly used predicates and high-level queries. See Table 11.1 for a complete list of these functions, most of whose names are clear indicators of their functionality. Note that they range from relatively low-level methods that return the position of an individual robot (`getTeamMembers`) to relatively high-level methods such as `NumOpponentsWithinDistance`. They include tactical considerations, such as whether `IAmClosestToBall`, as well as methods relative to our strategic roles (see Section 13.2.1), such as `GetDistanceFromSupporter`. Finally, methods such as `AmIInDefensiveZone` and `IsDefenderWellLocalized` provide a more abstract interface to the position estimates.

All of the interfaces described above were developed in the first year (2003). By the second year, we generalized and extended the Global Map into a structure (now called the "World State") that contained 178 different types of data, not including arrays of data (e.g. a history of accelerometer values over the last 13 cycles) and data structures that include many subfields (e.g. a joints and sensors data structure that contains all of the joint positions and sensor values of the AIBO). It also provided 132 functions that access these data. One notable addition to the WorldState object for 2004 was a distributed representation of the ball. Additional details on the WorldState object appear in Appendix C.1.

CHAPTER 12

Behaviors

This chapter describes the robot's soccer-playing behaviors. In our initial development, we had relatively little time to focus on behaviors, spending much more of our time building up the low-level modules such as walking, vision, and localization. As such, the behaviors described here are far from ideal. We later overhauled this component of our code base when we participated in subsequent competitions. Nonetheless, a detailed description of our initial year's behaviors is presented for the sake of completeness, and to illustrate what was possible in the time we had to work.

12.1 GOAL SCORING

One of the most important skills for a soccer-playing robot is the ability to score, at least on an empty goal. In this section, we describe our initial solution that was devised before the localization module was developed, followed by our eventual behavior that we used at RoboCup 2003.

12.1.1 Initial Solution

Once we had the initial movement and vision modules in place, we were in a position to "close the loop" by developing a very basic goal-scoring behavior. The goal was to test the various modules as they interacted with each other. Since neither the localization module (Chapter 8) nor the general architecture (Chapter 10) had been implemented by this time, this behavior was entirely reactive. In a project such as this one, it can be tempting to delay closing the loop until all components are in place. However, it is extremely important to forge ahead and obtain full autonomy as quickly as possible so as to solve the deceptively tricky integration issues and to shed light on what areas of the system are in most need of improvement. Our initial solution to goal-scoring is put forth here as an example of such early integration.

This goal scoring behavior, implemented as a Finite State Machine (FSM), assumes that the robot is placed at a point on the field such that the distance between the orange ball and the robot is not more than one-half the length of the field (i.e., the ball is at a distance where it can be seen by the robot). A point to note here is that this constraint could have been removed by incorporating a "random walk" sequence into the behavior. The robot first performs a three-layer head scan to determine if it can "see" the ball at its current position. If the ball is not in its visual field at this stage, the robot starts strafing (turning 360° about its current position) in search of the ball. In either case, the detection of a ball in a single visual frame causes the robot to stop and determine if the ball has actually been seen (noise in the image color segmentation can sometime cause false ball detections in high-level vision). Once the ball is in sight, the robot walk towards it by tracking the centroid of the ball with its head and moving its body in whatever direction the head points to. This walking state continues until either the ball is lost from the visual frame (in which case the robot goes back to searching for the ball) or the robot reaches a point sufficiently close to the ball, as determined by its neck angles at that point. The thresholds in the neck angles are set such that the robot stops with the ball right under its head. Next, the robot strafes around the ball with its head held at 0° tilt), searching for the offensive goal (blue or yellow depending on whether the robot is on the red team or the blue team). Once the goal is found, the robot checks to ensure that the ball is still under its nose and then tries to kick the ball into the goal. If the robot finds that it has lost the ball (it sometimes pushed it away accidentally while strafing), it goes back to searching for the ball.

This behavior, despite being extremely rudimentary, helped us understand the issues involved in the interaction/communication between modules. It also served to illustrate the importance of a good architecture in implementing complex behaviors. At the time of the American Open, this was the only goal-scoring behavior that we had implemented.

12.1.2 Incorporating Localization

When localization came into place, we replaced the above behavior using strafing and a single kick with a more complex behavior involving the chin pinch turn. In the new behavior, the decision about which kick to use is made according to the knowledge about where on the field the robot is and whether there are opponents nearby.

Figure 12.1 summarizes the kicking strategy used when no opponents are detected nearby. If the robot is on the offensive-half of the field and is not near any walls, it follows the natural strategy of turning toward the goal and then kicking the ball. On the quarter of the field nearest the offensive goal, the front-power kick is used rather than the fall-forward kick. This is because we believe the front-power kick to be more accurate than the fall-forward kick, although less powerful.

When the robot is in the defensive-half of the field, it kicks toward the far same-side corner (that is, if it is on the left-half of the field, it kicks toward the offensive-half left corner). The reasoning behind this was that when the ball is in the robot's defensive half, the most important thing is to clear the ball to the other half of the field. Since other robots are generally

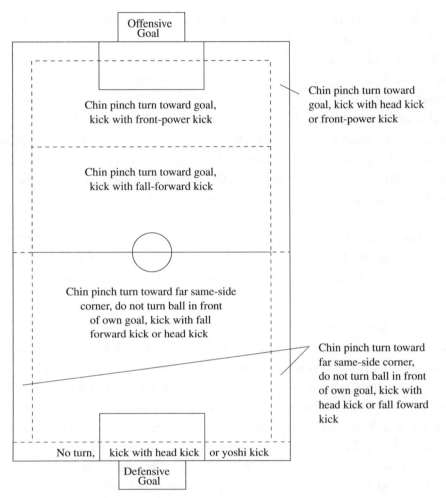

FIGURE 12.1: Kicking strategy when no nearby opponents are detected.

more likely to be in the center of the field, a good strategy for accomplishing this is to kick toward the outside so that the ball will on average be allowed to travel farther before its path is obstructed.

When the robot is near the wall and facing it, the head kick is typically used. This is chiefly because we want to use the chin pinch turn as little as possible when we are along the wall. The more the robot runs into the wall while moving, the larger the discrepancy becomes between the actual distance traveled and the information that odometry gives to localization. Because the FSM uses localization to determine when to switch from chin pinch turning to kicking, the longer the robot uses a chin pinch turn along a wall, the less likely it is to stop turning at the right time. So, it is typically a better strategy when very near a wall and facing it to head kick the ball along the wall rather than trying to turn with the ball to an exact angle and then kick with a more powerful kick.

Another situation where the head kick is used is when we would otherwise need to turn more than 180 degrees with the ball. This situation typically arises when the robot is in the defensive half and needs to avoid turning in a way that will pass the ball between it and its own goal. A 360-degree chin pinch turn takes approximately 5 s. Thus, given that many of our kicks take a small amount of time to prepare before hitting the ball away, chin pinch turning for more than 180 degrees carries the danger of putting us in violation of the 3-s holding rule. Therefore, in situations where we need to turn through some angle $\theta > 180$ degrees, we instead turn through $\theta - 80$ (or 180, if $\theta - 80 > 180$) degrees and then head kick in the appropriate direction.

If opponents are detected nearby, the robot simply kicks with the head kick in the direction of the goal (to the extent possible). The reasoning behind this choice is the same as the reasoning just described underlying the choice of the head kick near walls.

12.1.3 A Finite State Machine

Our behaviors are implemented by a Finite State Machine (FSM), wherein at any time the AIBO is in one of the finite number of states. The states correspond roughly to primitive behaviors, and the transitions between states depend on input from vision, localization, the global map, and joint angles. This section describes the FSM underlying our main goal-scoring behavior. As we developed our strategy more fully, this became the behavior of the attacker (see Section 13.2.1). The behaviors of the other two roles are discussed in Section 13.2.1 as well.

The main goal-scoring states are listed here. Note that the actions taken in these states are executed through the Movement Interface, and they are described in more detail in Section 5.2.2.

- Head Scan For Ball: This is the first of a few states designed to find the ball. While in this state, the robot stands in place scanning the field with its head. We use a two-layer head scan for this.

- Turning For Ball: This state corresponds to turning in place with the head in a fixed position (pointing ahead but tilted down slightly).

- Walking To Unseen Ball: This state is for when the robot does not see the ball itself, but one of its teammates communicates to it the ball's location. Then the robot tries to walk towards the ball. At the same time, it scans with its head to try to find the ball.

- Walking To Seen Ball: Here we see the ball and are walking towards it. During this state, the robot keeps its head pointed towards the ball and walks in the direction that its head is pointing. As the robot approaches the ball, it captures the ball by lowering its head right before transferring into the Chin Pinch Turn state.

- Chin Pinch Turn: This state pinches the ball between the robot's chin and the ground. It then turns with the ball to face the direction it is trying to kick.
- Kicking: While in this state, the robot is kicking the ball.
- Recover From Kick: Here the robot updates its knowledge of where the ball is and branches into another state. Both of these processes are influenced by which kick has just been performed.
- Stopped To See Ball: In this state, the robot is looking for the ball and has seen it, but still does not have a high enough confidence level that it is actually the ball (as opposed to a false positive from vision). To verify that the ball is there, the robot momentarily freezes in place. When the robot sees the ball for enough consecutive frames, it moves on to Walking To Seen Ball. If the robot fails to see the ball, it goes back to the state it was in last (where it was looking for the ball).

In order to navigate between these states, the FSM relies on a number of helper functions and variables that help it make state transition decisions. The most important of these are:

- BallLost: This Boolean variable denotes whether or not we are confident that we see the ball. This is a sticky version of whether or not high level vision is reporting a ball, meaning that if BallLost is true, it will become false only if the robot sees the ball (according to vision) for a number of consecutive frames. Similarly, a few consecutive frames of not seeing the ball are required for BallLost to become true.
- NearBall: This function is used when we are walking to the ball. It indicates when we are close enough to it to begin capturing the ball with a chin pinch motion. It is determined by a threshold value for the head's tilt angle.
- DetermineAndSetKick: This function is used when transitioning out of Walking To Seen Ball upon reaching the ball. It determines whether or not a chin pinch turn is necessary, what angle the robot should turn to with the ball before kicking, and which kick should be executed.

Finally, an overview of the rules that govern how the states transition into one another is given in Fig. 12.2.

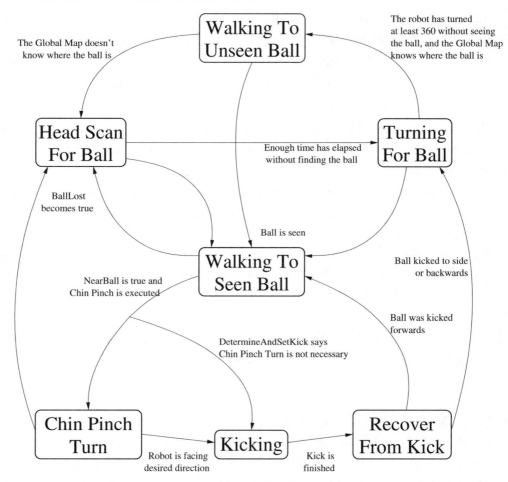

FIGURE 12.2: The finite state machine that governs the behavior of the attacking robot.

12.2   GOALIE

Like the initial goal-scoring behavior described in Section 12.1.1, our initial goalie behavior was developed without the benefit of localization and, as a result, is relatively crude. It is documented fully in our 2003 technical report [78]. This section details our RoboCup-2003 goalie behavior.

Once our goalie had the ability to determine its position on the field (localization), our primary strategy for the goalie focussed on staying between the ball and the goal. Given the large size of the goalie with respect to the goal, we adopted a fairly conservative strategy that kept the goalie in the goal most of the time.

Whenever the goalie saw the ball, it oriented itself such that it was pointed at the ball and situated between the ball and the goal. If the ball came within a certain distance of the goal, the goalie advanced towards the ball and attempted to clear it. After attempting to clear the ball, the goalie retreated back into the goal, walking backwards and looking for the ball. Any time the goalie saw the ball in a nonthreatening position, it oriented itself towards the ball and continued its current course of action.

Whenever the ball was in view, the goalie kept a history of ball positions and time estimates. This history allowed the goalie to approximate the velocity of the ball, which was useful in deciding when the goalie should "stretch out" to block a shot on the goal.

One interesting dilemma we encountered concerned the tradeoff between looking at the ball and looking around for landmarks. It seemed very possible that, given the goalie's size, if it could just stay between the ball and the goal it could to a fairly good job of preventing goals. However, this strategy depended on the goalie both being able to keep track of its own position and the ball's position. When we programmed the goalie to fixate on the ball, it was not able to see enough landmarks to maintain an accurate estimate of its own position. On the other hand, when the goalie focussed on the beacons in order to stay localized, it would often miss seeing the approaching ball. It proved to be very difficult to strike a balance between these two opposing forces.

CHAPTER 13

Coordination

This chapter describes our initial and eventual solutions to coordinating multiple soccer-playing robots.

13.1 DIBS

Our first efforts to make the players cooperate resulted directly from our attempts to play games with eight players. Every game would wind up with six robots crowded around the ball, wrestling for control. At this point, we only had two weeks before our first competition, and thus needed a solution that did not depend on localization, which was not yet functional. Our solution was a process we called *Dibs*.

13.1.1 Relevant Data

In developing Dibs, we tried to focus on determining both what data were available to us, and of that data, which were relevant. Because we did not have a coherent set of global maps at this point, any information from other robots would have to come directly into the Dibs system. As we created the system, it became more and more clear that the only thing we cared about was how far from the ball each robot was. Our first attempt simply transmitted the ball distance to every other robot. Each robot would then only go to the ball if its distance estimate was lower than that of every other robot.

13.1.2 Thrashing

Unfortunately, this first attempt did not work so well. First of all, the robots' perception of their distance to the ball was very heavily dependent on how much of the ball they could see, how the lights were reflecting off the surface of the ball, and how much of the ball was actually classified as "orange." This means that estimates of the ball's distance varied wildly from brain cycle to brain cycle, often by orders of magnitude in each direction. Secondly, even when estimates were fairly stable, a robot could think that it was the closest to the ball, start to step, and in the process move slightly backward, which would signal another robot to go for the ball. The other robot would begin to step, moving slightly backward at first, and the cycle would continue *ad infinitum*.

13.1.3 Stabilization

To correct these problems, we decided that re-evaluating which robot should go to the ball in each brain cycle was too much. Evaluating that frequently didn't give a robot the chance to actually step forward (this was before our walk was fully developed as well), so that its estimate of ball distance could decrease. However, we couldn't just take measurements every n brain cycles and throw away all the other information—we were strapped for information as it was, and we didn't want one noisy measurement to negatively affect the next n brain cycles of play. Our solution was to take an average of the measurements over a period of time, and instead only *transmit* them every n brain cycles.

13.1.4 Taking the Average

Because the vision is somewhat noisy (i.e., the robot sometimes sees the ball when it is not there, and sometimes doesn't see it when it is there), it didn't make sense just to take the raw mean of the estimates over the period of n brain cycles. We decided that unless the robot saw the ball for at least $\frac{n}{2}$ cycles in each period, it would report an essentially infinite distance to the ball. If it *did* see the ball enough, it would take all the noninfinite estimates in that "transmit cycle", discard some fixed number of the highest and lowest values (an attempt to clean up some of the noise), and then transmit the mean of the remaining values.

13.1.5 Aging

To prevent deadlock, we introduced an aging system into Dibs. Originally, if a robot had transmitted a very low estimate of distance to the ball, and then crashed or was removed from play, other robot would just remain watching the ball, because they would still have the other robot's estimate in their memory. Thus, at the end of each transmit cycle, we incremented the age of each other robot's estimate. When the age reached a predetermined cutoff (ten in our case), the estimate was discarded and set to the maximum value. In this way, other robots could then resume attacking the ball.

13.1.6 Calling the Ball

Another problem we ran into involved the "strafe" state. Once a robot had established "Dibs" on the ball, it would walk towards the ball while the other robots watch the ball closely. When the robot reached the ball, however, it would look up, in order to find the goal. While it was looking up, its ball estimates would all go to the maximum value, and other robots would resume attacking the ball. More often than not, this would result in a robot strafing to find the goal, while another robot of ours would come up and take the ball right out from under the nose of the first. Next, the second robot would start to strafe, and a large tangle of robots would result. To prevent this, we added functions called "callBall" and "relinquishBall." These functions merely set flags that made the robot start lying about its distance to the ball and stop lying, respectively. When lying about its distance to the ball, the robot would always report zero as its distance estimate. This way, whenever the robot entered the strafing state, it could effectively let the other robots know that even though it wasn't seeing the ball, they shouldn't go after it. The robot would then relinquish the ball at the beginning of most states, including when it had lost the ball and when it had just finished kicking the ball.

13.1.7 Support Distance

The system described so far worked pretty well in that it prevented more than one robot from going to the ball at once. However that was all it did. One robot might be going to the ball, but all the others would just stare at the ball, regardless of how far away they were. We determined that this was considerably sub-optimal, and that even if a robot is dribbling the ball down the field toward the enemy goal, if it were to lose the ball, it would be nice to have another robot nearby to recover, if possible. Thus, we introduced the concept of a "support distance." Originally set at half a meter, and then tuned to approximately a meter, the support distance was how close the robot would have to be to the ball before its lack of Dibs would prevent it from advancing further. While we only enjoyed limited overall success using the support distance technique, it was a marked improvement over ordinary Dibs.

13.1.8 Phasing out Dibs

Once localization was brought online, the need for multiple types of transmissions (which Dibs did not respect) and the desire to use localization data dictated a phasing out of Dibs. Because Dibs was so carefully tuned to the robots' playing style, cooperation actually worsened for quite a while before it improved after phasing out Dibs. However, as with many things, it needed to get worse before it could get better.

13.2 FINAL STRATEGY

This section describes the coordination strategy developed during the last week or so before RoboCup 2003. In particular, it takes advantage of both localization and global maps.

13.2.1 Roles

Our strategy uses a dynamic system of roles to coordinate the robots. In this system, each robot has one of three roles: *attacker*, *supporter*, and *defender*. The goalie does not participate in the role system. This section gives an overview of the ideas behind the roles. The following sections describe in more detail the supporter's and defender's behaviors and under what conditions the roles change.

The roles are dynamically assigned, in that at the start of each Brain cycle, a given robot re-evaluates its role based on its current role, its global map information, and other strategic information communicated to it by its teammates. The default allocation of roles is for there to be one defender and two attackers. Under certain circumstances, an attacker can become a supporter, but after some time it changes back into an attacker. It is also possible for the defender to switch roles with an attacker. There should always be exactly one defender and at least one attacker.

The differences between the roles manifest themselves in the robots' behaviors. Here is a summary of the differences between the behaviors effected by the different roles. The attacker's behavior is described in more detail in Section 12.1, and the supporter and defender behaviors are described more fully below.

- An attacker robot focuses exclusively on goal-scoring. That is, it tries to find the ball, move to it, and kick it towards the goal.

- The supporter's actions are based on a couple of goals. One is to stay out of the way of the attacker. This is based on the idea that one robot can score by itself more effectively than two robots both trying to score at the same time. Another goal is to be well placed so that if the attacker shoots the ball and it ricochets off the goalie or a wall, the supporter can then become the attacker and continue the attack.

- Our defender stays on the defensive-half of the field at all times. Its job is to wait for the ball to be on its half and then go to the ball and clear it back to the offensive side of the field.

13.2.2 Supporter Behavior

The supporter uses an omnidirectional walk to try to simultaneously face the ball and move to a *supporting post*. If it sees the ball, it keeps its head pointing towards the ball and tries to point its body in the same direction as its head. If it doesn't see the ball, it tries to turn towards its global map location of the ball and scans with its head to try to find it. It is very rare for there to be a supporter that has no idea where the ball is (i.e., while no robot sees the ball).

The location of the supporting post is a function of the position of the ball. For this we use a team-centric coordinate system where the edge of the field including the defensive goal line is the positive x-axis, the left edge of the field is the positive y-axis, and the units are millimeters. If the coordinates of the ball are (x, y), then the supporting post, (S_x, S_y), is given by

$$S_x = \begin{cases} 1150, & \text{if } x > 1450 \\ 1750, & \text{if } x \leq 1450 \end{cases} \qquad\qquad (13.1)$$

and

$$S_y = \min\left(\frac{y + 4200}{2}, 3800\right). \qquad\qquad (13.2)$$

13.2.3 Defender Behavior

The role of a defender in robot soccer is not much different from that in real soccer—to prevent the opponents from moving the ball anywhere near the goal it is defending and to try and kick the ball, when in its own half, towards a team member in the other half. We decided to go for a very conservative defender such that there is always one robot in our half defending the goal. At the same time, we wanted to ensure that under conditions where the defender is in a better position to function as the attacker, there is smooth switching of roles between the robots.

When a robot is assigned the defender role, its first action is to walk within a certain distance (approx. 200 mm) of a predefined defensive post that is roughly the center of the defensive half of the field. Once it gets within this distance of the defensive post, it either turns such that it faces the ball which is within its field of vision or it turns to face the point where it thinks the ball is based on the result of merging the estimates from other teammates in its global map (see Chapter 11). If it cannot see the ball and also does not receive any communication regarding the ball from other teammates (a rare occurrence), it starts searching for the ball once it gets to the defensive post. Even while it is walking to this post, if it sees the ball and finds, on the basis of its current world knowledge, that it is the closest to the ball, it starts walking to the ball. Once it gets to the ball it tries to kick the ball away from the defensive zone (the bottom three-fourths of the half of the field that it is defending). For the defender, we use a combination of the chin-pinch turn and the fall-forward kick (see Chapter 7), as it is the most powerful kick we have. While kicking, the defender always tries to angle the kick away from its own goal and towards one of the corners of the opposition. This strategy allows us to clear the ball in most instances and even takes it a long way into the other half thereby giving the attacker(s) (or attacker and supporter) a better chance of scoring a goal. According to the rules of the competition, none of the team members can enter the penalty box around their own goal. To accommodate this in the defender and in the other team members excluding the goalkeeper, we add a check that prevents the robot from entering the goal box and a "buffer" region around it. If the ball is within this region, the robot just tracks the ball and lets the goalkeeper take care of clearing the ball.

13.2.4 Dynamic Role Assignment

Our role assignment system has three main facets. One is a set of general rules that serve to maintain the status quo of there being exactly one defender and at least one attacker. Next are the rules that determine when one of the two attackers becomes a supporter and then when it switches back. The last set of rules orchestrates timely switches between the defender and an attacker.

General Rules

We label the three robots R_1, R_2, and R_3. Then the following rules influence R_1's choice of role. (The rules are the same for each robot; the labels are to distinguish which robot's role is being determined presently.)

- The default is for each robot to keep its current role. It will only change roles if a specific rule applies.

- If R_1 finds that it is "alone" in that it has not been receiving communication from other teammates for some time, it automatically assumes the role of an attacker.

- In most cases, communication works fine, and if neither R_2 nor R_3 is a defender, then R_1 will automatically become (or stay) a defender. This ensures that (under normal conditions) there will always be at least one defender. Ensuring that there is not more than one defender is taken care of in the section on attacker and defender switching.

- If R_1 is a supporter and so is R_2 or R_3, then R_1 will automatically become an attacker. This could happen accidentally if two supporters simultaneously decide to become supporters without enough time in between for the second one to be aware of the first's decision. In this case, this rule ensures that at least one of the supporters will immediately go back to being an attacker.

Attackers and Supporters

A number of considerations influence our mechanism for switching between attacker and supporter. One such consideration is that we want to prevent a robot from changing roles twice with very little time in between. This is because a robot that keeps changing roles very frequently behaves in a scattered manner and is unable to accomplish anything. To enforce this, we made the roles somewhat *sticky*. That is, for an attacker or supporter, there is an amount of time such that once the robot enters that role, it is unable to leave it until that much time has passed. Presently, the amount of time for an attacker is 2.5 s, and for a supporter it is 2 s. Notably, stickiness can easily be in conflict with the general rules listed above. In these cases, we give stickiness the highest priority. We also considered giving the general rules highest priority, and it is still not completely clear to us which system is better. Stickiness represents a form of *hysteresis*, which is a common and generally useful concept in robotics and control problems when it is important to prevent quick oscillations between different behaviors.

An important measure that we use to evaluate a robot's utility as an attacker is its *kick time*. This is an estimate of the amount of time it will take the robot in question to walk up to the ball, turn it towards the goal, and kick. Each robot calculates its own kick time and communicates it to the other robots as part of their communication of strategic information. The estimated amount of time to get to the ball is the estimated distance to the ball divided by the forward speed. The time to turn with the ball is determined by calculating the angle that the ball will have to be turned and dividing by the speed of the chin-pinch turn.

Consider the case where there are two attackers, A_1 and A_2. Once A_1's period of stickiness has expired, it will become a supporter precisely when *all* of the following conditions are met:

- A_1 and A_2 both see the ball. This helps to ensure the accuracy of the other information being used.

- The ball is in the offensive half, as well as both robots A_1 and A_2. Becoming a supporter is only useful when our team is on the attack.

- A_1 has a higher kick time than A_2. That is, A_2 is better suited to attack, so A_1 should become the supporter.

Once we have a supporter, S, and the role is no longer stuck, it will turn back into an attacker if *any* of the following conditions hold:

- S, the ball, or the attacker (A) go back into the defensive half.

- A and S both see the ball, and S's estimate of its distance from the ball is smaller than A's.

- A doesn't see the ball, and S's estimate of the ball's distance from it is less than some constant (presently 300 mm).

- S has been a supporter for longer than some constant amount of time (presently 12 s).

Attacker and Defender Switching

The following set of rules is used to allow the defender and an attacker to switch roles under appropriate circumstances.

- If a defender receives the information that there is another defender, it checks, using the global map data on the robots' distances to the ball, if it is a "better" defender (the one farthest from the ball). If so, it stays a defender. If not, it becomes an attacker.

- If a defender finds that there is no other defender, it still checks to see if the conditions are suitable for it to become an attacker. Here we test to see if the robot is closest to the ball and is in the section of the field that is on the top half on its side of the field. If it is, it sends a request to the attacker, asking to switch roles with it. Then, instead of becoming an attacker immediately, it waits for the attacker to receive the request. Once this happens, we end up with more than one defender in the team (see the rule mentioned below), and this is resolved using the condition mentioned above. More information on message types and communication can be found in Chapter 9.

- When an attacker receives a request from a defender to switch roles, it automatically accepts. It does not need to participate in the decision-making process because the defender had access to the same information as it did (as a result of the global maps) when it decided to switch. The attacker communicates its acceptance by simply becoming a defender. This is sufficient because the robots always communicate their roles to all of their teammates.

As mentioned above, our role system was developed quite hastily in the last week or so before competition. However, we feel that the system performs quite appropriately during games. The attacker/defender switches normally occur where they seem intuitively reasonable. The two attackers (with one becoming a supporter periodically), trying to score a goal, frequently look like a well-organized pair of teammates. Nonetheless, there are certainly some instances during the games where we can point to situations where a role change happened at an inopportune time, or where it seems like they should "know better" than to do what they just did. Finding viable solutions to problems like this can be strikingly difficult.

CHAPTER 14

Simulator

Debugging code and fine-tuning parameters are often cumbersome to perform on a physical robot. For example, the particle filtering implementation to localization that we used requires many parameters to be tuned. In addition, the robustness of the algorithm often works to mask errors, making them difficult to track down. For these reasons, we constructed a simulator that allows the robot's high-level modules to interact with a simulated environment through abstracted low-level sensors and actuators. The following sections describe this simulator in detail.

14.1 BASIC ARCHITECTURE

The simulator runs as a server, awaiting connections from client processes. One thread processes incoming UDP messages from clients. The other thread runs the simulation and sends UDP messages back to clients. The simulations run at 20 frames per second. The simulation begins after the first client connects.

All new robot clients start at the center of the field, facing the yellow goal. Each frame, the simulator uses the current action command for each client to update the robots' body positions and head angles according to a stochastic motion model. A stochastic sensor model is used to compute noisy observations for each of the robots. The observations are sent to the clients. The simulator displays the current state of the world and the internal state of the robots. The server then waits for update messages from the clients.

The client program is really just a wrapper around our world state code (the "Global Map" in Chapter 11) that handles all of the communication with the server. In addition, the client code specifies the agent's behavior using a simple mechanism, independent of the onboard behavior code (Chapter 12). This mechanism resembles a simple production system, and is capable of representing behaviors like following a figure-8 path around the field while performing a horizontal head scan.

14.2 SERVER MESSAGES

The client and server communicate using plain text message strings. The following message types are sent from the client to the server:

- **init** - Initialize contact with server.
- **param_walk** - Specify change in body movement.
- **move_head** - Specify change in head angle.
- **info** - Give server internal state info to be displayed by GUI.

The following message types are sent from the server to the client:

- **connect** - Acknowledge client initialization.
- **see** - Send observations for this time step.
- **sense** - Send joint feedback for this time step.
- **error** - Respond to incorrectly formatted client message.

The full grammar is specified in Appendix C.1.

14.3 SENSOR MODEL

The sensor model returns the distance and angle to fixed landmarks and the ball that fall into the view cone of the robot. The simulated robot is given the same view angle as the real AIBO. Currently, head tilt is ignored and assumed to be horizontal to the ground.

Gaussian noise is added to the distances and angles of each observation to model vision errors. The mean (bias) and variance of this noise can be varied by modifying parameters in the simulator. All observations are returned with a fixed confidence (currently 90%).

14.4 MOTION MODEL

The robot is moved by the motion model according to the last specified action command. The robot's body is moved according to the velocities specified in the param_walk command. The head pan is moved according to the move_head command. Unlike the param_walk command, move_head specifies an absolute position rather than a velocity. The simulator moves the head at a fixed speed to the specified pan angle and then holds it there.

The translational and rotational displacements of the robot are calculated by multiplying the velocities by the duration of a simulator step. Gaussian noise is added to the displacements and angles. The mean (bias) and variance of this noise can be changed by modifying parameters in the simulator. Joint feedback for the head pan angle is returned noise-free.

14.5 GRAPHICAL INTERFACE

As the simulator is running, the true position of each robot is displayed on a map of the field as a dark blue isosceles triangle with its apex pointing in the direction of its global orientation. A smaller triangle at the front of the robot, shows the robot's neck pan angle. In addition, the robot's pose estimate from localization is displayed as a light-blue triangle, and its particles are shown as white dots.

The user can reposition the robot by clicking on the robot's body with the left mouse button, dragging, then releasing. The orange ball can be moved around in this way as well. The robot's orientation can be changed by clicking on the robot's body with the right mouse button, dragging up and down, then releasing. Additional controls allow the user to pause the simulator and temporarily "blind" the robots by turning off sensory messages.

FIGURE 14.1: Screen shot of simulator. The robot's true position is in dark blue. Its estimated position is in light blue. The localization particles are shown in white. The ball estimate particles are shown in red.

CHAPTER 15

UT Assist

In addition to the simulator, which we used primarily for developing and testing localization, we developed a valuable tool to help us debug our robot behaviors and other modules. This tool, which we called UT Assist, allowed us to experience the world from the perspective of our AIBOs and monitor their internal states in real time.

15.1 GENERAL ARCHITECTURE

UT Assist consists of two pieces: a client and a server. The function of the client software, which is programmed in C++ and runs on an AIBO, is to queue and send data to the server. The server, which is programmed in Java and runs on a remote computer, is primarily concerned with collecting, displaying, and saving the data that it receives. We chose Java for the server because it put us on a relatively quick development cycle and gave us access to a rich library of pre-existing code. In particular, the ease with which Java handles networking and graphics made it an obvious candidate for this project.

Multiple clients can connect to one server. It is possible for more than one server to be active at once, provided that it does not listen on a port that is already taken by another service. All client–server communication takes place via TCP. The client software uses the default Open-R TCP endpoint interface, and the server software uses TCP networking classes described in the Java 2 API specification.

During each Brain cycle on the AIBO, many different pieces of code can attempt to send data messages to the server. If the client is not already sending data to the server, it will accept each request and place the specified data into a queue. If the client is busy sending data, it will reject the request to send data. At the end of each Brain cycle, if the client has some data in its queue, it will divide the data into fixed-length packets and start sending the data to the server. This method of processing data ensures that only data from the most recent Brain cycle will be sent to the server and avoids a "backlog" situation, in which the speed at which data is queued exceeds the speed at which it can be delivered to the server.

Each message that enters the queue in the client is uniquely identified by a one-byte ID field. From the perspective of the client, each message it receives is simply a group of bytes associated with a unique ID. None of the packet processing that the client performs upon the queue of messages depends on the actual data in the messages, which allows users to add new types of data messages without modifying the client.

When the data from the client reaches the server, it is reassembled into a queue of messages. Each message is then passed to the appropriate handler for that type of message. These different message types are discussed in the following section.

15.2   DEBUGGING DATA

One use of UT Assist is to extract debugging data from the AIBOs. The following sections describe the different types of data that can be viewed.

15.2.1 Visual Output

There are several different types of visual data that UT Assist can display, each of which allows the user to examine a different aspect of the AIBO's vision system. The different types of data are described as follows (see Fig. 15.1 for examples):

- Low- and high-resolution images – UT Assist can transfer full resolution images from the AIBO's camera (76,032 bytes in size) as well as smaller versions of the same image (4752 bytes) which lack the clarity of the high-resolution images but can be sent at a much faster rate. The small images can typically be transmitted and displayed in $frac13$ of a second, whereas the large images take 3–4 s to be displayed.

- Color-segmented images – The AIBO can send color-segmented images to UT Assist. The color-segmented images are the high-resolution images where each pixel has been

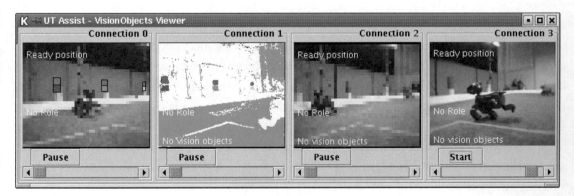

FIGURE 15.1: Several examples of visual data displayed in UT Assist. Part (a) contains a low-resolution image with vision objects overlayed on top. Part (b) shows a color-segmented image. Part (c) contains a low-resolution image, and (d) contains a high-resolution image.

classified by the AIBO's low-level vision as one of the several discrete colors (for more details, see Section 4.2).

- Vision objects – UT Assist can also display bounding boxes around objects that the AIBO's high-level vision software has recognized. UT Assist can overlay these bounding boxes on top of regular camera images, a technique that is useful for identifying possible bugs in the vision system.

15.2.2 Localization Output

UT Assist can also parse several types of debugging output which allow the user to examine aspects of the AIBO's internal model of the world. These data can be described as follows (see Fig. 15.2 for an example):

- Particle filtering – UT Assist can display the distribution of position/orientation particles that the AIBOs use to determine their position (for more on particles, see Chapter 8). In this case, it shows the distribution from running the *real* robots, as opposed to simulated versions as in Chapter 14.

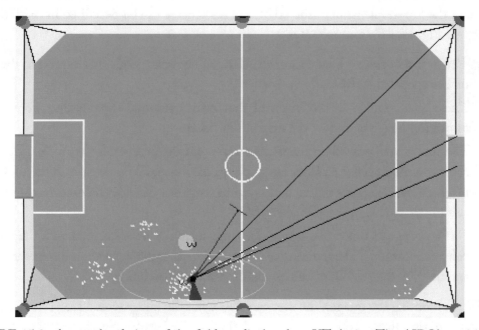

FIGURE 15.2: An overhead view of the field, as displayed on UT Assist. The AIBO's position and orientation are denoted by the small blue triangle. The white dots on the field represent the particle distribution that the AIBO uses to determine its position and orientation. Also shown are an estimate of error in the position (the blue oval), data from the IR sensor (the red line), and an estimate of the ball's position (the orange circle).

- Visible beacons – The AIBO can also send data about which beacons it currently sees, which can be displayed on the overhead map so that the user can quickly determine which beacons the AIBO can and cannot see.

- Position information – UT Assist can also display an AIBO's final estimate of position and orientation and the uncertainty associated with that estimate.

- Other objects – UT Assist can display an AIBO's estimate of where the ball currently is, as well as the location of any opponents that it sees.

15.2.3 Miscellaneous Output

- Infrared data – A small bar can be displayed in front of the AIBO's nose that indicates the current value of the infrared sensor.

- Text descriptions of state – Textual descriptions of the AIBO's current state and role can also be displayed (for more on roles, see Section 13.2.1).

15.3   VISION CALIBRATION

One of the chief benefits of UT Assist is the relatively seamless manner in which it can be used to calibrate the low-level vision of the AIBOs. This process can be described in the following manner (see Fig. 16.1 for examples):

1. The user requests an image from an AIBO. The server sends this request to the client on the specified AIBO.

2. The vision module in the Brain of that AIBO responds by sending with a high-resolution image back over the network to the server.

3. The image is displayed on the user's screen, and the user is allowed to "paint" various colors on the image (i.e., label the pixels on the image). The colors and the underlying pixels of the image are paired and saved so that they can later be used to compute the Intermediate (IM) cubes for the AIBO.

4. The user repeats this process, from step one, until she/he is satisfied with the resulting color calibration. Then the Master (M) cube is generated, loaded on the memory stick, and used by the AIBO for subsequent image classification.

While painting the images, the user can view what the image would look like had it been processed with the current color cube (NNr cube) using the Nearest Neighbor (NNr) rule. The user can also preview false-3D graphs of the YCbCr color space for each color. These represent the IM color cubes. The user can also see the M cube, generated by applying a NNr scheme on the NNr cube obtained by merging the IM cubes. For more details on color segmentation, see Section 4.2.

CHAPTER 16

Conclusion

The experiences and algorithms reported in this lecture comprise the birth of the UT Austin Villa legged-league robot team, as well as some of the subsequent developments. Thanks to a focussed effort over the course of 5 1/2 months, we were able to create an entirely new code base for the AIBOs from scratch and develop it to the point of competing in RoboCup 2003. The efforts were the basis of a great educational experience for a class of mostly graduate students, and then jump-started the Ph.D. research of many of them.

The results of our efforts in the RoboCup competitions between 2003 and 2005 are detailed in Appendix D.4. Though the competitions are motivating, exciting, and fun, the main goal of the UT Austin Villa robot legged soccer team is to produce cutting edge research results. Between 2003 and 2005, we have used the AIBO robots and our RoboCup team code infrastructure as the basis for novel research along several interesting directions. Here, the titles of some of these research papers are listed. Full details are available from our team web-page at http://www.cs.utexas.edu/~AustinVilla/?p=research.

Vision:

- *Real-Time Vision on a Mobile Robot Platform* [73].
- *Towards Illumination Invariance in the Legged League* [74].
- *Towards Eliminating Manual Color Calibration at RoboCup* [75].
- *Autonomous Color Learning on a Mobile Robot* [72].

Localization

- *Practical Vision-Based Monte Carlo Localization on a Legged Robot* [70].
- *Simultaneous Calibration of Action and Sensor Models on a Mobile Robot* [83].

Joint control

- *A Model-Based Approach to Robot Joint Control* [82].

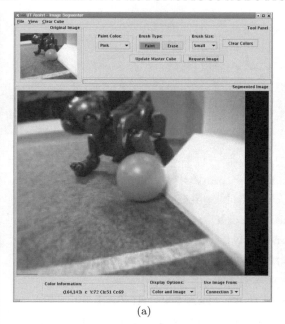

(a)

(b)

(c)

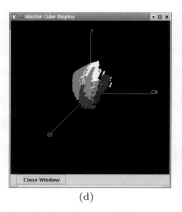

(d)

FIGURE 16.1: Color calibration for the AIBO using UT Assist. Part (a) shows the initial image as viewed in the UT Assist Image Segmenter. In (b), the user has started to classify (label) image pixels by painting various colors on them. Part (c) depicts the image after it has been classified, and (d) shows the distribution of colors in the three-dimensional YCbCr space, i.e., the final Master Cube.

Locomotion

- *Policy Gradient Reinforcement Learning for Fast Quadrupedal Locomotion* [44].
- *Machine Learning for Fast Quadrupedal Locomotion* [43].

Behavior learning

- *Learning Ball Acquisition on a Physical Robot* [23].

Robot surveillance

- *Continuous Area Sweeping: A Task Definition and Initial Approach* [1].

In summary, this lecture has been designed to provide a case study illustrating the challenges and benefits of working with a new robot platform for the purposes of education and research. Because the robot used was the Sony AIBO robot and the target task used was robot soccer, some of the details provided are specific to that robot and that task. But the intention is to convey general lessons that will continue to apply as new, more sophisticated, but still affordable robots continue to be rolled out. In addition, because the AIBO is a vision-based legged robot, many of the methods described are framed to apply to other robots that are vision-based, legged, or both. As robot technology continues to advance, more and more affordable robots that are suitable for class and research purposes are likely to share one or both of these characteristics.

Appendix A: Heuristics for the Vision Module

This appendix details the heuristics used in the vision module that are alluded to in the main text of Chapter 4.

A.1   REGION MERGING AND PRUNING PARAMETERS

After the initial color segmentation and runlength encoding, we attempt to merge the regions that correspond to the same object. This operation is really successful only if the associated color segmentation is good. But even then, choosing the heuristics used in the region merging process does affect the performance of the vision system as a whole. To a large extent, the thresholds were chosen based on the detailed experimentation with different numerical values.

- The first threshold to be decided is the extent to which two runlengths need to overlap (i.e., the number of pair of pixels, one from each runlength, that have the same horizontal coordinate) before we decide to merge them. We did try with several values corresponding to varying degrees of overlap. For example, if we set this number to be 1, we are asking for very little overlap between two runlengths (one below the other) while a number such as 10 would mean that a significant degree of overlap is expected between runlengths corresponding to the same object. We found that with our color segmentation, which does perform proper segmentation in most cases once it has been trained, we did not gain much in terms of accuracy by setting a high threshold. By setting a low threshold, on the other hand, we found that we rarely failed to merge regions corresponding to the same object. So, even though the low threshold did cause the generation of erroneous bounding boxes in some cases where the color segmentation did not do that well, we decided to go for a low threshold of overlap. We only look for an overlap of one to two pixels and use additional constraints (some of them are explained below) to remove the spurious blobs.

- Once the initial set of bounding boxes have been generated, we need to do some pruning to remove spurious blobs, particularly those caused due to the low merging threshold explained above. We realized that the numerous small blobs that were created need to be removed from consideration. We set a couple of thresholds here: the bounding box has to be at least 4 pixels wide along both the x- and y-axes, and each box needs to have at least 15 pixels to be considered any further. This works fine for the ball while for other colors we seem to need a lower (second) threshold to actually obtain the required performance (we reduce it to 9–10 pixels instead of the 15). This removes a lot of noisy estimates and those that survive are further pruned depending on the object being searched for (see subsequent appendices).

- Another heuristic that we include to remove spurious bounding boxes from further consideration is the density of the bounding boxes. The density of a bounding box is defined as:

$$\text{Density}_{\text{Bbox}} = \left(\frac{\text{no. of pixels of the color under consideration in Bbox}}{(\text{Bbox.lrx} - \text{Bbox.ulx}) \cdot (\text{Bbox.lry} - \text{Bbox.uly})} \right) \quad \text{(A.1)}$$

We determined experimentally that a minimum density requirement of ≥ 0.4 ensures that we remove most of the spurious bounding boxes generated due to boundary effect, lighting variations, etc., and at the same time the really significant blobs are retained for further consideration.

A.2   TILT-ANGLE TEST

The tilt-angle test is one of the most widely used heuristics to remove spurious blobs that are generated due to shadowing and reflectance problems. These are generally the blobs that appear to be floating in the sky or are on the field and hence cannot represent objects of interest. The motivation for this approach is the fact that the objects of interest rarely appear above the horizon in the image (they do not appear in the field or on the opponents either). This test is mainly used only in the case where the robot's head is tilted down by an angle less than $25°$ because with the head tilted much lower, the objects may appear above the horizon too.

In this method, we use the known camera rotation parameters and the bounding boxes obtained in the image under consideration to obtain the *compensated* tilt angle at which the object would have been observed, with respect to the robot's camera frame of reference, if the robot's head had been at its reference (base) position (i.e., zero tilt, pan, and roll).

$$\text{Compensated}_{\text{tilt}}(\text{radians}) = \arctan \left(\frac{\text{ImgCenter}_y - \text{BboxCentroid}_y}{\text{FocalPixConstant}} \right) + \text{RoboCamTilt}$$

$$\text{(A.2)}$$

where

1. ImgCenter_y: This is the center of the image plane along the y-axis (x varies from 0 to 175 while y varies from 0 to 143) given by $\lfloor \frac{143}{2} \rfloor = 71$.

2. BboxCentroid$_y$: This is the y-coordinate of the centroid of the bounding box under consideration, in the image plane.

3. FocalPixConstant: This is a constant of the robot's camera system, given by *ImageResolution* · *CameraFocalLength* = 72 · 2.18.

4. RoboCamTilt: This is the basic camera tilt of the robot when the image is observed.

This Compensated$_{tilt}$ can then be used in conjunction with experimentally determined thresholds to remove the spurious blobs from further consideration in the object-detection phase. For example, in the case of the ball, we could easily set a threshold and say that the compensated tilt should not be greater than 1° for all valid balls. See subsequent appendices for details on individual thresholds for various objects on the field.

A.3   CIRCLE METHOD

The orange ball is probably the most important object that needs to be detected on the field. We need a good estimate of the ball bounding box as it determines its size and hence its distance from the robot. This information is very important for planning the game strategy. But in most cases the ball is partially occluded by the other robots and the objects on the field or by the fact that only part of the ball is within the robot's visual field.

Once the ball bounding box has been determined, we find three points on the circumference of the ball by scanning along three lines, one each along the top, bottom, and center line of the bounding box. Each of these points is found by searching for the orange-to-any-color and any-color-to-orange transition. Of course, we do run into problems when the basic color segmentation of the ball is not perfect and we find colors such as yellow and red on sections of the ball. But in most cases this method gives a good estimate of the ball size and hence distance (an error of ±10 cm in distance).

Given the three points, we can determine the equation of the circle that passes through these points and hence obtain the center and radius of the circle that describes the ball and provides an estimate of the ball size even for partially occluded balls. Consider the case where we find three points in the image plane, say $P_1 = (x_1, y_1)$, $P_2 = (x_2, y_2)$, and $P_3 = (x_3, y_3)$ (see Fig. A.1). Since all these points lie on the circle that describes the ball, we can then write:

$$(x - x_1)^2 + (y - y_1)^2 = 0$$
$$(x - x_2)^2 + (y - y_2)^2 = 0$$
$$(x - x_3)^2 + (y - y_3)^2 = 0.$$

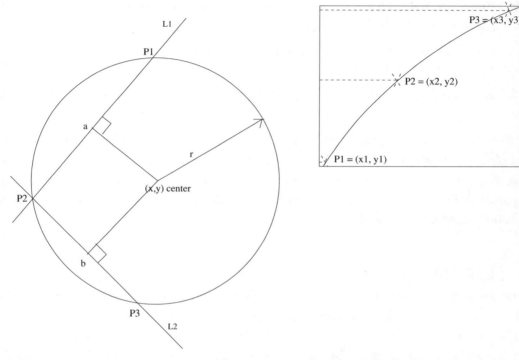

FIGURE A.1: Given three point P_1, P_2, P_3, we need to find the equation of the circle passing through them.

Through the pair of points P_1, P_2 and P_2, P_3, we can form two lines L_1, L_2. The equations of the two lines are

$$y_{L_1} = m_{L_1} \cdot (x - x_1) + y_1 \qquad (A.3)$$
$$y_{L_2} = m_{L_2} \cdot (x - x_2) + y_2, \qquad (A.4)$$

where

$$m_{L_1} = \left(\frac{y_2 - y_1}{x_2 - x_1}\right), \qquad m_{L_2} = \left(\frac{y_3 - y_2}{x_3 - x_2}\right) \qquad (A.5)$$

The center of the circle is the point of intersection of the perpendiculars to lines L_1, L_2, passing through the mid points of segments $P_1 - P_2$, (a) and $P_2 - P_3$, (b). The equations of the perpendiculars are obtained as

$$y'_{c-a} = \left(\frac{-1}{m_{L_1}}\right) \cdot \left(x - \frac{x_1 + x_2}{2}\right) + \frac{y_1 + y_2}{2} \qquad (A.6)$$

$$y'_{c-b} = \left(\frac{-1}{m_{L_2}}\right) \cdot \left(x - \frac{x_2 + x_3}{2}\right) + \frac{y_2 + y_3}{2}. \qquad (A.7)$$

Solving for x gives

$$x = \frac{m_{L_1} \cdot m_{L_2} \cdot (y_1 - y_3) + (m_{L_2} - m_{L_1}) \cdot x_2 + (m_{L_2} \cdot x_1 - m_{L_1} \cdot x_3)}{2 \cdot (m_{L_2} - m_{L_1})} \tag{A.8}$$

$$y = \frac{(x_1 - x_3) + (m_{L_1} - m_{L_2}) \cdot y_2 + (m_{L_2} \cdot y_1 - m_{L_1} \cdot y_3)}{2 \cdot (m_{L_2} - m_{L_1})}. \tag{A.9}$$

This gives the radius of the circle as

$$r = \sqrt{(x - x_1)^2 + (y - y_1)^2}. \tag{A.10}$$

This gives us all the parameters we need to get the size of the ball.

A.4   BEACON PARAMETERS

In the case of beacons, several parameters need to be set, based on experimental values.

1. The two bounding boxes that form the two sections of the beacon are allowed to form a *legal* beacon *if* the number of pixels and number of runlengths in each section are at least one-half of that in the other section.

2. None of the two regions must be 'too big' in comparison with the other. 'Too big' refers to cases where the x- and/or y-coordinate ranges of one section are greater than 3–3.5 times the corresponding ranges of the other bounding box. If 'ul' and 'lr' denote the upper-left and lower-right corners of the bounding box, then, for the x-coordinates, we would consider either of the following equations as the sufficient condition for rejection of the formation of a beacon with these two blobs:
    $(b1.\text{lr}x - b1.\text{ul}x + 1) \geq 3 \cdot (b2.\text{lr}x - b2.\text{ul}x + 1)$
    $(b2.\text{lr}x - b2.\text{ul}x + 1) \geq 3 \cdot (b1.\text{lr}x - b1.\text{ul}x + 1).$
    Similarly, for the y-coordinates
    $(b1.\text{lr}y - b1.\text{ul}y + 1) \geq 3 \cdot (b2.\text{lr}y - b2.\text{ul}y + 1)$
    $(b2.\text{lr}y - b2.\text{ul}y + 1) \geq 3 \cdot (b1.\text{lr}y - b1.\text{ul}y + 1).$
    The runregion size checks (see Appendix A.1) help ensure the removal of beacons that are too small.

3. The x-coordinate of the centroid of each section must lie within the x range values of that of the other section, i.e.,
    $b1.\text{ul}x \leq b2_{\text{centroid}x} \leq b1.\text{lr}x$
    $b2.\text{ul}x \leq b1_{\text{centroid}x} \leq b2.\text{lr}x.$

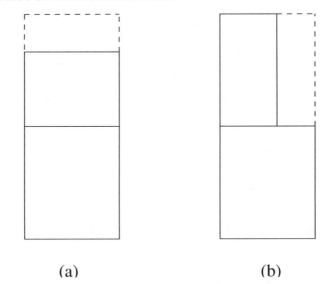

(a)   (b)

FIGURE A.2: This figure shows the basic beacon size extension to compensate for partial occlusions. Case (a) is an example of vertical extension while case (b) depicts an example of horizontal extension.

4. The distance between sections (in number of pixels) is also used to decide whether two bounding boxes can be combined to form a beacon. In our case, this threshold is 3 pixels.

5. *Size extension:* In the case of the beacons, similar to the ball (though not to that large a degree), occlusion by other robots can cause part of the beacon to be 'chopped off'. On the basis of our region merging and beacon-region-matching size constraints, we can still detect beacons with a fair amount of occlusion. But once the beacons have been determined, we extend the size of the beacon such that its dimensions correspond to that of the larger beacon subsection determined (see Fig. A.2).

6. *Beacon likelihoods:* After the beacon dimensions have been suitably extended, we use the estimated size and the known aspect ratio in the actual environment to arrive at a likelihood measure for our estimation. The beacon has a *Height* : *Width* :: 2 : 1 aspect ratio. In the ideal case, we would expect a similar ratio in the image also. So we compare the aspect ratio in the image with the desired aspect ratio and this provides us with an initial estimate of the likelihood of the estimate. Whenever there are multiple occurrences of the same beacon, this value is used as a criterion and the 'most-likely' beacon values are retained for further calculations. Further, since the occurrence of false positives is a greater problem (for localization) than the case where we miss some beacon, we only use beacons with a likelihood ≥ 0.6 for localization computations.

A.5   GOAL PARAMETERS

The parameters/heuristics for the goal were selected experimentally and they affect the performance of the robot with respect to the goal detection. Some of them are as enumerated below.

1. We desire to be able to detect the goals accurately both when they are at a distance and when they are really close to the robot and in each case the image captured by the robot (of the goal) and hence the bounding boxes formed are significantly different. So we use almost the same criteria as with other objects (runlengths, pixels, density, aspect ratio, etc.) but specify ranges. The ranges are chosen appropriately; we first use strict constraints and search for ideal goals that are clearly visible and are at close range (this gets assigned a high likelihood—see next point), but if we fail to do so, we relax the constraints and try again to find the goals. In addition, we have slightly different parameters tuned for the yellow and blue goals because the identification of the two goals differs based on the lighting conditions, robot camera settings, etc.

    - *Runlengths:* at a minimum, we require 10–14.

    - *Number of pixels:* high values 3000–4000+, low values 200–400+.

    - *Aspect ratio:* length/width : 1.1–1.3 (at least) but not more than 2.5–3.0.

    - *Density:* at least 0.5+.

2. *Tilt-angle test:* In the case of the goals, we do not want them to be either too high or too low with respect to the horizon. So we apply the same tilt angle heuristic (Appendix A.2) but with two thresholds. In our case, these angles are in the range: $7°$ to $10°$ (high) and $-11°$ to $-8°$ (low).

3. We also observed conditions wherein a yellow blob appears in the ball, when some portions of the orange ball have nonuniform illumination and/or reflectance properties (this should not be considered as the goal). Another heuristic is therefore calculated to prevent this: the position of the blob centroid with respect to the ball centroid. If the size of the goal is smaller and the centroid lies somewhere on the ball, we are likely to reject this estimate of the goal.

4. *Goal likelihood:* After the goal dimensions have been determined, we use the estimated size and the known aspect ratio in the actual environment to arrive at a likelihood measure for our estimation. The goals have a *Height : Width* :: 1 : 2 aspect ratio. In the ideal case, we would expect a similar ratio in the image also. So we compare the aspect ratio in the image with the desired aspect ratio and this provides us with an initial estimate of the likelihood of the estimate. Further, we only use goals with a likelihood ≥ 0.6 for localization computations. In fact, we use the goal edges and not the goals for

localization and they are assumed to have the same likelihood as the goal they belong to.

A.6   BALL PARAMETERS

In the case of the ball, again several of the tests are the same as those in the case of the goals and/or beacons but the parameters are different, these are determined experimentally.

1. There are some general constraints:
   - *Density:* at least 0.5+ and *aspect ratio:* length/width : 0.7–1.3 (strict).
   - Relax *aspect ratio* constraints but with

     (a) *Runlengths:* at a minimum we require 10–14 and,
     (b) *Number of pixels:* high values 1000–1400+, low values 200–400+ or
     (c) *Size:* extends to $\frac{1}{3}$ of the length or height of the image frame.

2. *Tilt test:* The ball cannot 'float in the air'. We apply the tilt-angle heuristic with an upper threshold of 1° to 5°.

3. *Circle method:* In this case, we ignore the ball size generated by this method if it is too small (2 pixels), too large (85 pixels) or much smaller than the 'uncompensated' size that existed before the circle method was applied.

4. *Ball likelihood:* Here, we choose a simple method to assign the likelihood: assign a high likelihood (0.75–0.9 depending on the size of the ball) if the circle method generates a valid ball size and assign a low likelihood (0.5–0.7) if the circle method fails and we are forced to accept the initial estimate.

A.7   OPPONENT DETECTION PARAMETERS

This section provides some of the parameters used in the opponent detection process. As mentioned in the vision module (Section 4.4), the height of the blob is used to arrive at an estimate of the distance to the opponent and its bearing with respect to the robot, by the same approach used with other objects in the image frame. Some sample thresholds

1. For the basic detection of a blob as a candidate opponent blob, we use the constraints on runlengths, number of pixels, etc., which decide the tradeoff between accuracy and the maximum distance at which the opponents can be recognized.
   - *Pixel threshold:* We set a threshold of 150–300 pixels.
   - *Runlengths:* We require around 10.

2. *Tilt test:* The opponents cannot float much above the ground and cannot appear much below the horizon (in the ground) with the robot's head not being tilted much. The threshold values here are 1° (high) and −10° (low).

3. *Merging in vision:* This is similar to the region merging process. Two blobs that are reasonably close in the visual frame are merged if the interblob distance is less in the range of 20–30 pixels. Varying the threshold varies the opponent detection 'resolution', i.e., how far two opponents have to be to be recognized as two different robots.

A.8   OPPONENT BLOB LIKELIHOOD CALCULATION

We use an extremely simple approach to determine the likelihood of the opponents found in the image. This is done by comparing the properties of the opponent blob with the 'ideal' values (those that correspond to the actual presence of an opponent) determined by the experimentation (some are listed in the previous appendix).

- A member of the opponent list that has more than 450–500 pixels and more than 10 runlengths is given a very high likelihood (0.9+).

- For other members, we make the likelihood proportional to the maximum based on the number of pixels.

- Blobs that have low probability are not accepted in the list of opponents. Also, these get eliminated very easily during merging with estimates from other teammates.

A.9   COORDINATE TRANSFORMS

Consider the case where we want to transform from the local coordinate frame to the global coordinate frame. Figure A.3 shows the basic coordinate system arrangement. To find the position of any point (x_l, y_l) given in the local coordinate system (x, y), with respect to the global coordinate system (X, Y), we use the knowledge of the fact that the local coordinate system has its origin at (p_x, p_y) and is oriented at an angle θ with respect to the global coordinate frame.

$$\begin{pmatrix} X_g \\ Y_g \\ 1 \end{pmatrix} = \begin{pmatrix} \cos(\theta) & -\sin(\theta) & p_x \\ \sin(\theta) & \cos(\theta) & p_y \\ 0 & 0 & 1 \end{pmatrix} \begin{pmatrix} x_l \\ y_l \\ 1 \end{pmatrix} \quad (A.11)$$

By a similar matrix transform, we can move from the global coordinate frame to the local coordinate frame.

$$\begin{pmatrix} x_l \\ y_l \\ 1 \end{pmatrix} = \begin{pmatrix} \cos(\theta) & -\sin(\theta) & -p_x \cos(\theta) - p_y \sin(\theta) \\ \sin(\theta) & \cos(\theta) & p_x \sin(\theta) - p_y \cos(\theta) \\ 0 & 0 & 1 \end{pmatrix} \begin{pmatrix} X_g \\ Y_g \\ 1 \end{pmatrix} \quad (A.12)$$

These are the equations that we refer to whenever we speak about transforming from the local to global coordinates or vice versa. For more details on coordinate transforms in 2D and/or 3D, see [67].

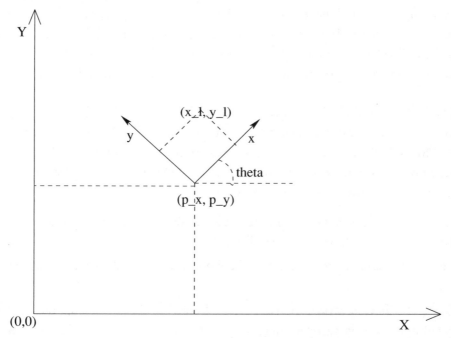

FIGURE A.3: This figure shows the basic global and local coordinate systems.

A.9.1 Walking Parameters

This section lists and describes all 20 parameters of our AIBO walk as described in Section 5.1. The units for most of the parameters are distances which are in terms of leg-link length, as discussed in Section 5.1.2. Exceptions are noted below.

- Forward-step distance: How far forward the foot should move from its home position in one step.

- Side-step distance: How far sideways the foot should move from its home position in one step.

- Turn-step distance: How far each half step should be for turning.

- Front-shoulder height: How high from the ground the robot's front legs' J_1 and J_2 joints should be.

- Back-shoulder height: How high from the ground the robot's back legs' J_1 and J_2 joints should be.

- Ground fraction: What fraction of a step time the robot's foot is on the ground. (The rest of the time is spent with the foot in the air, making a half-ellipse.) Between 0 and 1. Has no unit.

- Front-left y-offset: How far out in the y-direction the robot's front-left leg should be when it's in its home position.

- Front-right y-offset: How far out in the y-direction the robot's front-right leg should be when it's in its home position.

- Back-left y-offset: How far out in the y-direction the robot's back-left leg should be when it's in its home position.

- Back-right y-offset: How far out in the y-direction the robot's back-right leg should be when it's in its home position.

- Front-left x-offset: How far out in the x-direction the robot's front-left leg should be when it's in its home position.

- Front-right x-offset: How far out in the x-direction the robot's front-right leg should be when it's in its home position.

- Back-left x-offset: How far out in the x-direction the robot's back-left leg should be when it's in its home position.

- Back-right x-offset: How far out in the x-direction the robot's back-right leg should be when it's in its home position.

- Front clearance: How far up the front legs should be lifted off the ground at the peak point of the half-ellipse.

- Back clearance: How far up the back legs should be lifted off the ground at the peak point of the half-ellipse.

- Direction_fwd: Whether the robot should move forwards or backwards. Either 1 or -1. Has no unit.

- Direction_side: Whether the robot should move right or left. Either 1 or -1. Has no unit.

- Direction_turn: Whether the robot should turn towards its right or its left. Either 1 or -1. Has no unit.

- Moving_max_counter: Number of Open-R frames one step takes. Greater than 1. Has no unit.

Appendix B: Kicks

Kicking has been an area of continual refinement for our team. The general kick architecture that was a part of our initial development is presented in Chapter 7. Six of the kicks developed in the first year are summarized here.

B.1   INITIAL KICK

As discussed in Chapter 7, refined kicking was not an early priority in our team development. However, we did need at least one example to complete the development and integration of the other modules. Thus, we created a "first kick" early on to address the needs of the other modules as they developed and created other kicks much later to expand our strategic capabilities.

We decided to model our first kick after what seemed to be the predominant goal-scoring kick from previous RoboCup competitions. During the kick, the robot raises its two front legs up and drops them onto the sides of the ball. The force of the falling legs propels the ball forward. Our first kick, called the "front power kick" tried to achieve this effect.

We wanted our front power kick to transition from any walk without prematurely tapping the ball out of the way. Thus, we started the kick in a "broadbase" position in which the robot's torso is on the ground with its legs spread out to the side. If the robot were to transition into the front power kick from a standing position, the robot would drop to the ground while pulling its legs away from the ball. From this broadbase position, the robot then moves its front legs together to center the ball. After the ball has been centered, the robot moves its front legs up above its head and then quickly drops the front legs onto the sides of the ball, kicking the ball forward.

We found that the kick moves the ball relatively straight ahead for a distance of up to 3 m. However, we noticed that the robot's front legs would miss the ball if the ball were within 3 cm of the robot's chest. We resolved this issue by using the robot's mouth to push the ball slightly forward before dropping its legs on the ball.

B.2   HEAD KICK

After many of our modules had been integrated, the need arose for a kick in a nonforward direction. Inspired by previous RoboCup teams, decided that the head could be used to kick the ball to the left or to the right. During the head kick, the robot first leans in the direction opposite of the direction it intends to kick the ball. The robot then moves its front leg (left leg when kicking left, right leg when kicking right) out of the way. Finally, the robot leans in the direction of the kick as the head turns to kick the ball.

The head kick moves the ball almost due left (or right) a distance of up to 0.5 m. We discovered that the head kick was especially useful when the ball was close to the edge of the field. The robot could walk to the ball, head kick the ball along the wall, and almost immediately continue walking, whereas the front power kick frequently kicked the ball against the wall, effectively moving the ball very little, if at all.

B.3   CHEST-PUSH KICK

The creation of the head kick informed us that the robot could enter and exit a kick much faster when the kick occurred with the robot in a standing position. We thus created the chest-push kick in hopes that its execution would be much faster than that of the front power kick. During the chest-push kick, the robot quickly leans its chest into the ball. This occurs while the robot remains in a standing position.

To create the kick, we first isolated the kick from the walk. The following table shows the critical action for the chest push kick. In these tables, each value of Δt is listed in the row of the *Pose* that ends the corresponding *Move*.

TABLE B.1: Chest Push Kick Critical Action

	j_1	j_2	j_3	j_4	j_5	j_6	j_7	j_8	j_9	j_{10}	j_{11}	j_{12}	j_{13}	j_{14}	j_{15}	j_{16}	Δt
$Pose_1$	−12	30	91	−12	30	91	−70	45	104	−70	45	104	0	0	0	0	64
$Pose_2$	−120	90	145	−120	90	145	120	25	125	120	25	125	0	0	0	0	1
$Pose_3$	−12	30	91	−12	30	91	−30	6	104	−30	6	104	0	0	0	0	64

TABLE B.2: Chest-Push Kick Initial Action And Critical Action

	j_1	j_2	j_3	j_4	j_5	j_6	j_7	j_8	j_9	j_{10}	j_{11}	j_{12}	j_{13}	j_{14}	j_{15}	j_{16}	Δt
$Pose_s$	−12	30	91	−12	30	91	−30	6	104	−30	6	104	0	0	0	0	64
$Pose_1$	−12	30	91	−12	30	91	−70	45	104	−70	45	104	0	0	0	0	64
$Pose_2$	−120	90	145	−120	90	145	120	25	125	120	25	125	0	0	0	0	1
$Pose_3$	−12	30	91	−12	30	91	−30	6	104	−30	6	104	0	0	0	0	64

We then integrated the walk with the kick. Testing revealed that the robot successfully kicked the ball 55% of the time and fell over after 55% of the successful kicks. Since $(Pose_y, Pose_1, \Delta t)$ added unwanted momentum to the critical action, we created an initial action to precede the critical action. $\{(Pose_y, Pose_s, 64), (Pose_s, Pose_1, 64)\}$ does not lend unwanted momentum to the critical action. Testing revealed that the robot now successfully kicked the ball 100% of the time. The following table shows the initial action with the critical action.

Since the critical action did not add unwanted momentum that hindered the robot's ability to resume its baseline motion, there was no need to create a final action.

We found that the chest-push kick moves the ball relatively straight ahead. It is also very fast. However, the distance the ball travels after the chest-push kick is significantly smaller than the distance the ball travels after the front power kick. Thus, we decided against using the chest-push kick instead of the front power kick during the game play.

B.4 ARMS TOGETHER KICK

After creating kicks geared toward scoring goals, we realized that we needed a kick for the goalie to block the ball from entering its goal. Deciding that speed and coverage area were more important than the direction of the kick, we created the arms-together kick. During the arms-together kick, the robot first drops into broadbase position mentioned in Section B.1. The robot then swings its front-left leg inward. After that, the robot swings its front-right leg inward as it swings its front-left leg back out. The arms-together kick proved successful at quickly propelling the ball away from the goal.

B.5 FALL-FORWARD KICK

After attending the American Open, we saw a need for a forward-direction kick more powerful than the front power kick. Inspired by a kick used by the CMPack team from Carnegie Mellon, we created the fall-forward kick. The fall-forward kick makes use of the forward momentum of the robot as it falls from standing position to lying position. Since the kick begins in a standing position, the robot can quickly transit from the walk to the kick. However, since the kick ends in a lying position, the robot does not transit from the kick back to the walk as quickly.

TABLE B.3: Fall-Forward Kick Critical Action																
	j_1	j_2	j_3	j_4	j_5	j_6	j_7	j_8	j_9	j_{10}	j_{11}	j_{12}	j_{13}	j_{14}	j_{15}	j_{16} Δt
$Pose_1$	−5	0	20	−5	0	20	−35	6	75	−35	6	75	45	−90	0	0 32
$Pose_2$	−100	23	0	−100	23	0	100	6	75	100	6	75	45	−90	0	0 32

TABLE B.4: Fall-Forward Kick Critical Action And $\{(Pose_2, Pose_g, 32)\}$

	j_1	j_2	j_3	j_4	j_5	j_6	j_7	j_8	j_9	j_{10}	j_{11}	j_{12}	j_{13}	j_{14}	j_{15}	j_{16}	Δt
$Pose_1$	−5	0	20	−5	0	20	−35	6	75	−35	6	75	45	−90	0	0	32
$Pose_2$	−100	23	0	−100	23	0	100	6	75	100	6	75	45	−90	0	0	32
$Pose_g$	90	90	0	90	90	0	100	6	75	100	6	75	45	−90	0	0	32

We first isolated the kick from the walk. The following table shows the critical action.

We then integrated the walk with the kick. There was no need to create an initial action because any momentum resulting from $(Pose_y, Pose_1, 32)$ was in the forward direction (the same direction we wanted the robot to fall). However, testing revealed that $(Pose_2, Pose_z, \Delta t)$ caused the robot to fall forward on its face every time. Although the robot successfully kicked the ball, the robot could not immediately resume walking. In this situation, the robot had to wait for its fall detection to trigger and tell it to get up before resuming the walk. The get up routine triggered by fall detection was very slow. Thus, we found a $Pose_g$ such that $\{(Pose_2, Pose_g, 32), (Pose_g, Pose_z, \Delta t)\}$ does not hinder the robot's ability to resume walking. The following table shows the critical action with $Move(Pose_2, Pose_g, 32)$.

From the observation, it is noted that transitioning from $Pose_2$ directly to $Pose_g$ is not ideal. The robot would fall over 25% of the time during $(Pose_2, Pose_g, 32)$. Thus, we added $Pose_w$ to precede $Pose_g$ in the final action. Afterwards, the robot no longer fell over when transitioning from the kick to the walk. The following table shows the entire finely controlled action, consisting of the critical action and the final action.

The fall-forward kick executed quickly and potentially moved the ball the entire distance of the field (4.2 m). Unfortunately, the fall-forward kick did not reliably propel the ball directly

TABLE B.5: Fall-forward Kick Critical Action and Final Action

	j_1	j_2	j_3	j_4	j_5	j_6	j_7	j_8	j_9	j_{10}	j_{11}	j_{12}	j_{13}	j_{14}	j_{15}	j_{16}	Δt
$Pose_1$	−5	0	20	−5	0	20	−35	6	75	−35	6	75	45	−90	0	0	32
$Pose_2$	−100	23	0	−100	23	0	100	6	75	100	6	75	45	−90	0	0	32
$Pose_w$	−100	90	0	−100	90	0	100	6	75	100	6	75	45	−90	0	0	32
$Pose_g$	90	90	0	90	90	0	100	6	75	100	6	75	45	−90	0	0	32

forward. Thus, in the game play, we used the fall-forward kick in the defensive-half of the field and used the front power kick for more reliable goal scoring in the offensive-half of the field.

One unexpected side-effect of adding $Pose_g$ to the end of the fall-forward kick was that the outstretched legs in $Pose_g$ added additional ball coverage. A ball that the fall-forward action missed because it was not located around the robot's chest would actually be propelled forward if the ball was just in front of one of the front legs. Thus, the fall-forward kick, which moves the ball away much farther than the arms-together kick, also became our primary goalie block.

B.6   BACK KICK

Games at the American Open also inspired us to create the back kick. During the back kick, the robot launches its body over the ball and kicks the ball out from behind it. The back kick ideally works well in situations when the robots are crowded together around the ball. However, because the back kick is still somewhat unreliable, the behavior used for RoboCup games only executes a back kick in very specific circumstances, which in practice occur rarely. (See Section 12.1.2 for details.)

Appendix C: TCPGateway

As described in Chapter 9, the RoboCup competitions provided a module called TCPGateway, that was used in the RoboCup competitions and that abstracted away most of the low-level networking, providing a standard Open-R interface in its place. This brief section describes our interface with that module.

The TCPGateway configuration files insert two network addresses and ports in the middle of an Open-R subject/observer relationship. This creates the following situation:

- Instead of sending data directly to the intended observer, the subject on the initiating robot sends data to a TCPGateway observer.

- The TCPGateway module on the initiating robot has a specific connection on a unique port for data flowing in that direction, and sends the data from the subject over that connection to the PC.

- The PC, which has been configured to map data from one incoming port to one outgoing port, sends the subject's data out to the receiving robot on a specific port.

- The TCPGateway module on the receiving robot processes the data that it receives on this port and sends the data to the intended observer.

All of the mappings described above were defined in two files on each robot (CONNECT.CFG and ROBOTGW.CFG) and in two files on the PC (CONNECT.CFG and HOSTGW.CFG).

C.1 EXTENSION TO WORLD STATE IN 2004

All of the interfaces described above were developed in the first year of our project (2003). By the second year, we generalized and extended the Global Map into a structure (now called the "World State") that contained 178 different types of data, not including arrays of data (e.g. a history of accelerometer values over the last 13 cycles) and data structures that include many subfields (e.g. a joints and sensors data structure that contains all of the joint positions and sensor values of the AIBO). Some examples of this data include

- Time information: the current time, the amount of time spent processing this Brain cycle, the amount of time spent processing the last Brain cycle.

- The current score of the game.

- Information about the state of each teammate: their positions, roles, ideas about the location of the ball.

- Which teammates are still "alive" (i.e. not crashed).

- The position of the AIBO on the field.

- The position of the ball and a history of where it has been.

- What objects the AIBO currently sees.

- Matrices that describe the current body position of the AIBO.

- A history of which electrostatic sensors have been activated, and for how long.

- All of the communication data, including messages from our teammates and from the game controller.

Our WorldState object also provided 132 functions that access these data. Some of these functions return the data in its original form, but most of them process the data to provide a higher-level view of that data. For example, there are a number of ways we can access the position of the AIBO on the field:

- Position in absolute field coordinates.

- Position in team-relative field coordinates.

- Position relative to some object (e.g. own goal or opponent's goal).

- Fuzzy team-relative position (e.g. on right side of the field, near own goal).

- Role-sensitive position (e.g. is there an attacker in the opponent's half of the field?).

- Information about the location of the ball with respect to objects on the field (e.g. ourselves, teammates, our own goal).

- Raw angles and distances from various objects on the field to other objects on the field.

One notable addition to the WorldState object for 2004 was a distributed representation of the ball. Instead of only having a single position estimate for the ball, this distributed representation used a variable number (in our case, 20) of particles, where each particle represented one possible location for the ball. A probability was associated with each particle that indicated how confident we were in that estimate of the ball's position. While this approach was originally inspired by particle filtering (an algorithm used by our localization algorithm—see Chapter 8), it was implemented as a much simpler algorithm. In order to get the behavior that we wanted out of these "ball particles", we dramatically reduced the effect of the probability update step, instead choosing to update the particles almost entirely by reseeding. Particles were reseeded (i.e. one particle was added to the filter, replacing a low-probability particle) whenever the

AIBO saw the ball for two consecutive frames or whenever one of its teammates was able to see the ball and communicate that information to the AIBO.

One way that we used the information from these ball particles was by averaging the particles in one angular dimension that centered around the AIBO, so that we could use behaviors that required that we pick a single angle for the ball (for example, in order to look at the ball we would want to know what angle to turn the head to). The method that we used to extract a single ball *position* from the ball particles was to cluster the particles according to their (x, y) coordinates on the field, in a manner similar to that of our localization algorithm (for more information about this clustering algorithm, see the localization section in our 2003 technical report [78]). Having a distributed representation of the ball allowed the AIBOs to deal gracefully with conflicting reports of the ball's position caused by vision and localization errors.

Some of the information in the WorldState is shared between the AIBOs, using the communication framework described in Chapter 9. One of the first steps in each Brain cycle involves the AIBO checking for information from other teammates and inserting that information into its own WorldState. At the end of each Brain cycle, each AIBO sends some of its own information to its teammates. The following information is transmitted between AIBOs:

- Ping/pong messages—used to determine which AIBOs are alive and connected to the network.

- Ball messages—the position of the ball, whether the AIBO currently sees that ball or not, and strategy information about whether the AIBO is currently allowed to approach the ball or not.

- Position messages—the position of the AIBO on the field as well as its current role.

- Command messages—information that instructs the AIBO to change its current role and tells it whether it can approach the ball or not (as elaborated in Chapter 13).

Appendix D: Simulator Message Grammar

This appendix contains a complete grammar for the simulator messages sent from the client and server in our simulator that we used for the development of localization algorithms (see Chapter 14).

⟨CLIENT-MSG⟩→ ⟨INIT-MSG⟩ | ⟨PARAM-WALK-MSG⟩ | ⟨MOVE-HEAD-MSG⟩ | ⟨INFO-MSG⟩

⟨INIT-MSG⟩→ (init)

⟨PARAM-WALK-MSG⟩→ (param_walk ⟨INTEGER⟩ ⟨INTEGER⟩ ⟨INTEGER⟩)

⟨MOVE-HEAD-MSG⟩→ (move_head ⟨INTEGER⟩ ⟨INTEGER⟩)

⟨INFO-MSG⟩→ (info ⟨INFO⟩*)

⟨INFO⟩→ ⟨ESTIMATE-INFO⟩ | ⟨PARTICLE-INFO⟩ | ⟨BALL-PARTICLE-INFO⟩

⟨ESTIMATE-INFO⟩→ (e ⟨INTEGER⟩ ⟨INTEGER⟩ ⟨INTEGER⟩ ⟨INTEGER⟩ ⟨INTEGER⟩ ⟨INTEGER⟩)

⟨PARTICLE-INFO⟩→ (p ⟨INTEGER⟩ ⟨INTEGER⟩ ⟨INTEGER⟩ ⟨INTEGER⟩)

⟨BALL-PARTICLE-INFO⟩→ (b ⟨INTEGER⟩ ⟨INTEGER⟩ ⟨INTEGER⟩)

⟨SERVER-MSG⟩→ ⟨CONNECT-MSG⟩ | ⟨SENSE-MSG⟩ | ⟨SEE-MSG⟩ | ⟨ERROR-MSG⟩

⟨CONNECT-MSG⟩→ (connect)

⟨SENSE-MSG⟩→ (sense ⟨INTEGER⟩ ⟨SENSATION⟩*)

⟨SENSATION⟩→ ⟨HEAD-SENSE⟩

⟨HEAD-SENSE⟩→ (h ⟨INTEGER⟩ ⟨INTEGER⟩)

⟨SEE-MSG⟩→ (see ⟨INTEGER⟩ ⟨OBSERVATION⟩*)

⟨OBSERVATION⟩→ ⟨BALL-OBS⟩ | ⟨LANDMARK-OBS⟩

⟨BALL-OBS⟩→ (b ⟨INTEGER⟩ ⟨INTEGER⟩ ⟨INTEGER⟩)

⟨LANDMARK-OBS⟩→ (l ⟨INTEGER⟩ ⟨INTEGER⟩ ⟨INTEGER⟩ ⟨INTEGER⟩)

⟨ERROR-MSG⟩→ (error ⟨STRING⟩)

⟨INTEGER⟩→ 0 | [\-]? [1-9] [0-9]*

⟨STRING⟩→ [a-zA-Z\-]$^{+}$

D.1   CLIENT ACTION MESSAGES

The following messages are sent by the client to change its action.

- **(param_walk** ∂x ∂y $\partial \theta$**)**
  Set the robot's translational velocity to $\langle \partial x, \partial y \rangle$ in mm/s and its rotational velocity to $\partial \theta$ in degrees per second.

- **(move_head** *pan tilt***)**
  Move robot's head to angles *pan* and *tilt* in degrees.

D.2   CLIENT INFO MESSAGES

The following strings are sent by the client in **info** messages to supply internal state information to the server.

- **(e** *x y* θ *certainty pan tilt***)**
  The current pose estimate produced by the localization is $\langle x, y, \theta \rangle$ in mm and degrees. Its pose estimate confidence is *certainty*%. Its estimate of its head joint angles are *pan* and *tilt* in degrees.

- **(p** *x y* θ *p***)**
  One of the robot's localization particles has pose $\langle x, y, \theta \rangle$ in mm and degrees and probability *p*%.

- **(b** *x y p***)**
  One of the robot's ball particles has position $\langle x, y \rangle$ in mm and degrees and probability *p*%.

D.3   SIMULATED SENSATION MESSAGES

The simulator's **sense** messages, which act to emulate the robot's sensors and joint feedback are formatted as follows.

- **(sense** *cycle sensations. . .* **)**
  The *sensations* were sensed at time step *cycle*.

The following sensation strings are sent as part of **sense** messages. Currently, only one type of sensation is supported.

- **(h** *pan tilt***)**
  The robot's head joint feedback returns angles *pan* and *tilt* in degrees.

D.4   SIMULATED OBSERVATION MESSAGES

The simulator's **see** messages, which act to emulate the robot's vision module are formatted as follows.

- **(see** *cycle observations...* **)**
  The *observations* were made at time step *cycle*.

  The following observation strings are sent as part of **see** messages.

- **(b** *d* α *certainty*)
  The robot observes a ball at distance *d* in mm at heading α in degrees with confidence *certainty*%.

- **(l** *id* *d* α *certainty*)
  The robot observes a landmark #*id* at distance *d* in mm at heading α in degrees with confidence *certainty*%.

Appendix E: Competition Results

In the RoboCup initiative, periodic competitions create fixed deadlines that serve as important motivators. Our initial goal was to have a team ready to enter in the First American Open Competition in April 2003. We then proceeded to qualify for and enter the Seventh International RoboCup Competition in July of 2003 and also these same events in subsequent years. This appendix describes our results and experiences at those events.

E.1   AMERICAN OPEN 2003

The First American Open RoboCup Competition was held in Pittsburgh, PA, from April 30th to May 4th, 2003. Eight teams competed in the four-legged league, four of which, including us, were teams competing in a four-legged league RoboCup event for the first time. The eight teams were divided into two groups of four for a round robin competition to determine the top two teams which would advance to the semi-finals. The teams in our group were from the University of Pennsylvania, Georgia Institute of Technology (two veteran teams), and Tec de Monterrey, Mexico (another new team). At this competition, we used our initial behaviors for both the goal-scoring player (see Section 12.1.1) and the goalie [78]. The results of our three games are shown in Table E.1.

On the day before the competition, we arranged for a practice game against the Metrobots, a new team from three schools in the New York Metropolitan area: Columbia, Rutgers, and Brooklyn College. The game was meant as an initial test of our behaviors. In particular, we used SplineWalk in the first half and ParamWalk in the second half. After going down 3–0 in the first half, the game ended 4–1 in favor of the Metrobots.

Playing in this practice game was very valuable to us. Immediately afterwards, we drew up the following priority list:

1. Don't see the ball off the field. There was a parquet floor and wooden doors in the room. The robot appeared to often see them as orange and try to walk off the field.

2. Get localization into the goalie. At the time, we were using the initial goalie solution, and the goalie often found itself stuck at the corner.

3. Start with a set play. When we had the kickoff, the other team often got to the ball before we were ready to kick it. We decided to try reflexively kicking the ball and having another robot walk immediately to the target of the kick.

TABLE E.1: The Scores of Our Three Games at the American Open

OPPONENT	SCORE (US–THEM)	NOTES
Monterrey	1–1	Lost the penalty shootout 1–2
Penn	0–6	
Georgia Tech	0–2	

4. Get a sideways head kick in. Whereas our initial goal-scoring behavior strafed around the ball to look for the goal, the other team's robots often walked to the ball and immediately flicked it with their heads. That provided them with a big advantage.

5. Move faster. Most other teams in the tournament used CMU's walk from their 2002 code. Those teams moved almost twice as fast as ours.

6. Kick more reliably and quickly. It was often clear that our robots were making good decisions. However, their attempted kicks tended to fail or take too long to set up.

Some of these priorities—particularly the last two—were clearly too long-term to implement in time for the competition. But they were important lessons.

Overnight, we added some preliminary localization capabilities to the goalie so that it could get back into the goal more quickly. We also developed an initial set play and made some progress toward kicking more quickly when the ball was near the goal (though not on the rest of the field) by just walking forward when seeing the orange ball directly in front of the goal.

Our set play for use on offensive kickoffs involved two robots. The first one, placed directly in front of the ball, simply moved straight to the ball and kicked it directly forward—without taking any time to localize first. We tended to place the robot such that it would kick the ball towards an offensive corner of the field. The second robot, which was placed near the corresponding center beacon, reflexively walked forwards until either seeing the ball or timing out after moving for about half a meter.

On defensive kickoffs, our set play was simply to instruct one of the robots to walk forwards as quickly as possible. We placed that robot so that it was facing the ball initially.

In the first game, we tied Tec do Monterrey 1–1 in a largely uneventful game. In the penalty shootout, we lost 2–1. Though we only attained limited success, given the time we had to develop our team, we were quite happy to score a goal and earn a point in the competition (1 point for a shootout loss).

After some network problems for both teams, the 2nd game against Penn ended up as a 6–0 loss. Penn, the eventual runner-up at RoboCup 2003, had a very unique and powerful kick in which the robot turns while its arm is stuck out. Despite the lopsided score, we did observe some positive things. The goalie made some good plays and successfully localized on the fly a couple of times. Our new set play worked a little bit, too.

Our last game was against Georgia Tech, the eventual runner-up of the competition, and we only lost 2–0. The goalie looked good again—it was certainly the player with the most action overall.

One of the biggest take-home lessons from this competition was that although our robots appeared to make intelligent decisions, they had no hope in the competition unless they could walk and kick as quickly as the other robots. We briefly considered moving to the CMU 2002 walking and kicking code as we proceeded with our development toward RoboCup 2003. However, in the end, we decided to stick with our decision to create our entire code base from scratch.

Over the next two months, we continued developing the code outside of the class context. Many of the routines described in this lecture were developed over the course of those two months. In particular, we did succeed in creating a faster walk (Chapter 5), we got localization working (Chapter 8), we developed many more kicks (Chapter 7), and we completely reworked the strategy (Chapter 12), all as described herein. During this time, we played frequent practice games with two full teams of robots, which helped us immensely with regards to benchmarking our progress and exploring the space of possible strategies.

By the end of June, we were much more prepared for RoboCup 2003 than we had been for the American Open. Of course we expected the competition to be tougher as well.

E.2   ROBOCUP 2003

The Seventh International RoboCup Competition was held in Padova, Italy, from July 2nd to 9th, 2003. Twenty-four teams competed in the four-legged league, eight of which, including us, were teams competing at the international event for the first time. The 24 teams were divided into four groups of six for a round robin competition to determine the top two teams which would advance to the quarter-finals. The teams in our group were the German Team from University of Bremen, TU Darmstadt, Humboldt University, and University of Dortmund, all in Germany; ASURA from Kyushu Institute of Technology and Fukuoka Institute of Technology in Japan; UPennalizers from the University of Pennsylvania; Essex Rovers from the University of Essex in the UK; and UTS Unleashed! from the University of Technology at Sydney. Essex ended up being unable to compete and dropped out of the competition. The results of our four games are shown in Table E.2.

TABLE E.2: The Scores of Our Four Official Games at RoboCup

OPPONENT	SCORE (US–THEM)
UTS Unleashed!	1–7
German Team	0–9
UPennalizers	0-6
ASURA	1-4

Like at the American Open, we made sure to arrange some practice games in Italy. The results are shown in Table E.3. Our first test match was against CMU, the defending champions. We ended up with an encouraging 2–2 tie, but it was only their first test match as well, with some things still clearly not working correctly yet. Overall, we were fairly satisfied with our performance.

In our first "official" practice match (organized by the league chairs), we played against the University of Washington team (3rd place at the American Open) and lost 1–0. It was a fairly even game. They scored with 10 s left in the half (the game was just one-half). They also had one other clear chance that they kicked the wrong way. In this game, it became apparent that we had introduced some bugs while tuning the code since the day before. For example, the fall detection was no longer working. We also noticed that our goalie often turned around

TABLE E.3: The Scores of Our Six Unofficial Games at RoboCup

OPPONENT	SCORE (US–THEM)
CMU	2–2
U. Washington	0–1
Team Sweden	3–0
U. Washington	0–4
Metrobots	3–1
Team Upsalla	4–0

to face its own goal in order to position itself. It was in that position when it was scored on. The other noticeable problem was that our robots had a blind spot when looking for the ball: there were times when we should have gotten possession of the ball but did not see it.

Nonetheless, we remained happy with our performance. The role switching was working well, and our robots were as fast to the ball in general as any other team's. We had the ball down in UW's end of the field frequently. We just couldn't get any good shots off.

We won our 2nd and last "official" practice game against team Sweden 3–0. They had some problems with their goalie, so our first two goals were essentially on an empty net. We were hoping to test some changes to the goalie in this game, but the ball was in their end most of the time, so our goalie didn't get tested much.

Next, we played another informal practice match against the University of Washington team and lost 4–0. This game appeared to be much worse for us than the previous one against them, so we decided to undo some of the changes we had made on site. Although it is always tempting to keep trying to improve the team at the last minute, it is also risky. This is an important lesson about competitions that has been learned many times and is still often ignored!

In our first official game, we played the other new team in our group, UTS Unleashed! and lost 7–1. Our impression was that the score was not reflective of the overall play: there wasn't anything noticeably wrong with our code. UTS Unleashed! was just much more efficient at converting their chances. Playing in this game exposed our goalie's weakness with regards to being unable to both remain localized and see the ball at the same time.

Next, we played the top-seeded team in our group, the German team, and lost 9–0. Again, our opponent did not appear so much better than us, but the small things made a big difference in terms of goals. Our general feeling was one of pride at having caught up with the other teams in terms of many of the low-level skills such as fast walking, kicking, localization, etc. But we just didn't have the time to meld those into quite as tuned a soccer strategy as those of the other teams. One highlight of this game was a wonderful save by our goalie in which it swatted the ball away with a dive at the last second and then followed the ball out to clear it away.

In our next game, we played the eventual tournament runner up, UPennalizers, and lost 6–0.

In place of our canceled game against Essex, we decided to have a rematch from the American Open against the Metrobots. This time we met with a much different result, winning 3–1.

We played our last official game against Asura, the winner of the Japan Open. It was 1–1 at halftime, but we ended up losing 4–1. Still, we continued to be happy with the way the team looked in general. The ball was in our offensive end of the field a fair amount. We were just less able to score when we had chances, and our goalie continued to be a weak link on defense.

Finally, we played one last practice match against Team Upsalla from Sweden and won 4–0.

Based on all of our practice matches, we seemed to be one of the better new teams at the competition. We were in a particularly hard group, but we were able to compete at a reasonable level with even the best teams (despite the lopsided scores).

E.3   CHALLENGE EVENTS 2003

In addition to the actual games, there was a parallel "challenge event" competition in which teams programmed robots to do three special-purpose tasks:

1. Locating and shooting a black and white (instead of an orange) ball;
2. Localizing without the aid of the standard six-colored field markers; and
3. Navigating from one end of the field to the other as quickly as possible without running into any obstacles.

Given how much effort we needed to put in just to create an operational team in time for the competition, we did not focus very much attention on the challenges until we arrived in Italy. Nonetheless, we were able to do quite well, which we take as a testament to the strengths of our overall team design.

On the first challenge, we finished in the middle of the pack. Our robot did not succeed at getting all the way to the black and white ball (only eight teams succeeded at that), but of all the teams that did not get to the ball, our robot was one of the closest to it, which was the tie-breaking scoring criterion. Our rank in this event was 12th.

In this challenge event, we used our normal vision system with a change in high-level vision for ball detection (see Section 4.4 for details on objection recognition). The black and white ball appears almost fully white from a distance, i.e., in cases where we can see the entire ball, and the algorithm first searched for such blobs whose bounding boxes had the required aspect ratio (1:1). In other cases in which the ball is partially occluded, the ball was visualized as being made up of black and white blobs, and the idea was to group similar sized blobs that were significantly close to each other (a threshold determined by the experimentation). This required us to also train on the black and white ball when building up the color cube (see Section 4.2 for details on color segmentation). This approach worked well in our lab, but on the day of the challenge we did not have the properly trained color cube on the robot, which resulted in the robot not being able to see the ball well enough to go to it.

In the localization challenge, the robot was given five previously unknown points on the field and had to navigate precisely to them without the help of the beacons. Our robot used the goals to localize initially, and then relied largely on odometry to find the points. Our robot successfully navigated to only one of the five points, but the large majority of teams failed to do even that. Our score was sufficient to rank us 5th place in this event. Unfortunately, we were disqualified on a technicality. We had initially programmed the robot with the wrong coordinate system (a mere sign change). Rather than running the robot toward mirror images of the actual target points, we decided to fix the code and accept the disqualification.

Finally, in challenge 3, the robot was to move from one side of the field to the other as quickly as possible without touching any of the seven stationary robots placed in previously unknown positions. Our robot used an attraction and repulsion approach which pulled it toward the target location but repelled it from any observed obstacle. The resulting vector forces were added to determine the instantaneous direction of motion for the robot. Since speed was of the essence, our robot would switch to our fastest gait (ParamWalk) when no obstacles were in sight. A slower gait that allowed omnidirectional movement (SplineWalk) was used for all other movement.

Our robot was one of only four to make it all the way across the field without touching an obstacle, and it did so in only 63.38 s. The German team succeeded in just 35.76 s, but the next closest competitor, ARAIBO, took 104.45 s. Thus, we ranked 2nd in this event.

Officially, we finished 13th in the challenge events. However, the unofficial results, which did not take into account our disqualification in event 2, nor one for the University of Washington, placed UT Austin Villa at the fourth place. Given that 16 of the 24 RoboCup teams were returning after having competed before, and several of them had spent more effort preparing for the challenges than we had, we were quite proud of this result and are encouraged by what it indicates about the general robustness of our code base.

E.4   U.S. OPEN 2004

The Second U.S. Open RoboCup Competition was held in New Orleans, LA, from April 24th to 27th, 2004.[1] Eight teams competed in the four-legged league, and were divided into two groups of four for a round robin competition to determine the top two teams which would advance to the semifinals. The three other teams in our group were from Georgia Institute of Technology, Dortmund University, and Instituto Tecnologico Autonoma de Mexico (ITAM), from Mexico. After finishing in second place in the group, we advanced to the semifinals against a team from the University of Pennsylvania, and eventually the tournaments third place game against Dortmund. The results of

[1]http://www.cs.uno.edu/~usopen04/.

TABLE E.4: The Scores of Our Five Games at the U.S. Open

OPPONENT	SCORE (US–THEM)	NOTES
ITAM	8–0	
Dortmund	2–4	
Georgia Tech	7–0	
Penn	2–3	Semifinal
Dortmund	4–3	

our three games are shown in Table E.4. Links to videos from these games are available at http://www.cs.utexas.edu/~AustinVilla/?p=competitions/US_open_2004.

Our first game against ITAM earned us our first official win in any RoboCup competition. The score was 3–0 after the first half and 8–0 at the end. The attacker's adjustment mechanism designed to shoot around the goalie (Section 12.1) was directly responsible for at least one of the goals (and several in later games).

Both ITAM and Georgia Tech were still using the smaller and slower ERS-210 robots, which put them at a considerable disadvantage. Dortmund, like us, was using the ERS-7 robots. Although leading the team from Dortmund 2–1 at halftime, we ended up losing 4–2. But by beating Georgia Tech, we still finished 2nd in the group and advanced to the semi-finals against Penn, who won the other group.

Our main weakness in the game against Dortmund was that our goalie often lost track of the ball when it was nearby. We focused our energy on improving the goalie, eventually converging on the behavior described in Section 12.2, which worked considerably better.

Nonetheless, the changes weren't quite enough to beat a good Penn team. Again we were winning in the first half and it was tied 2–2 at halftime, but Penn managed to score the only goal in the second half to advance to the finals.

Happily, our new and improved goalie made a difference in the third place game where we won the rematch against Dortmund by a score of 4–3. Thus, we won the 3rd place trophy at the competition!

We came away from the competition looking forward towards the RoboCup 2004 competition two months later. Our main priorities for improvement were related to improving the localization including the ability to actively localize, adding more powerful and varied kicks, and more sophisticated coordination schemes.

E.5   ROBOCUP 2004

The Eighth International RoboCup Competition was held in Lisbon, Portugal, from June 28th to July 5th, 2004.[2] Twenty-four teams competed in the four-legged league and were divided into four groups of six for a round robin competition to determine the top two teams which would advance to the quarterfinals. The teams in our group were the ARAIBO from The University of Tokyo and Chuo University in Japan; UChile from the Universidad de Chile; Les 3 Mousquetaires from the Versailles Robotics Lab; and Penn. Wright Eagle from USTC in China was also scheduled to be in the group, but was unable to attend. After finishing 2nd in our group, we qualified for a quarter-final match-up against the NuBots, the University of Newcastle in Australia. The results of our five games are shown in Table E.5. Links to videos from these games are available at `http://www.cs.utexas.edu/~AustinVilla/?p=competitions/roboCup_2004`.

TABLE E.5: The Scores of Our Five Games at RoboCup

OPPONENT	SCORE (US–THEM)	NOTES
Les 3 Mousquetaires	10–0	
ARAIBO	6–0	
Penn	3–3	
Chile	10–0	
NuBots	5–6	Quarterfinal

In this pool, Les 3 Mousquetaires and Chile were both using ERS-210 robots, while the other teams were all using ERS-7s. In many of the first round games, the communication among robots and the game controller was not working very well (for all teams), and thus reduced performance on all sides. In particular, in the game against Penn, the robots had to be started manually and were unable to reliably switch roles. In that game, we were winning 2–0, but again saw Penn come back, this time to tie the game 3–3.

This result left us in a tie with Penn for first place in the group going into the final game. Since the tiebreaker was goal difference, we needed to beat Chile by two goals more than Penn beat ARAIBO in the respective last games in order to be the group's top seed. Penn proceeded to beat ARAIBO 9–0, leaving us in the unfortunate position of needing to score

[2]`http://www.robocup2004.pt/`.

11 goals against Chile to tie Penn, or 12 to pass them. The 10–0 victory left us in the second place and playing the top seed of another group in the quarterfinals, the NuBots.

In the quarterfinal, the network was working fine, so we got to see our robots at full speed. We scored first twice to go up 2–0. But by halftime, we were down 4–2. In the 2nd half, we came back to tie 4–4, then went down 5–4, then tied again. With 2 minutes left, the NuBots scored again to make it 6–5, which is how it ended. It was an exciting match, and demonstrated that our team was competitive with some of the best teams in the competition (Penn and the NuBots both lost in the semifinals, though). In the end, we were quite pleased with our team's performance.

E.6   U.S. OPEN 2005

The Third U.S. Open RoboCup Competition was held in Atlanta, GA, from May 7th to 10th, 2005. Eight teams competed in the four-legged league, and were divided into two groups of four for a round robin competition to determine the top two teams which would advance to the semifinals. The three other teams in our group were from Georgia Institute of Technology, Spelman College, and Columbia University/CUNY. After finishing at first place in the group, we advanced to the semifinals against a team from the University of Pennsylvania, and eventually the tournament's third-place game against Columbia. The results of our four games are shown in Table E.6. Links to videos from these games are available at http://www.cs.utexas.edu/~AustinVilla/?p=competitions/US_open_2005.

Overall, our performance was quite strong, giving up only a single goal and scoring 22. Unfortunately, the one goal against was in the semifinal against Penn in a very close game. CMU eventually beat Penn 2–1 in the final (in overtime). But as an indication of how evenly matched the three top teams are, we beat CMU in an exhibition match 2–1. We also played an

TABLE E.6: The Scores of Our Five Games at the U.S. Open

OPPONENT	SCORE (US–THEM)	NOTES
Georgia Tech	4–0	
Columbia/CUNY	3–0	
Spelman	7–0	
Penn	0–1	Semifinal
Columbia	8–0	

exhibition match against a team from Dortmund, the winner of the 2005 German Open, and lost 2–0.

E.7   ROBOCUP 2005

The Ninth International RoboCup Competition was held in Osaka, Japan, from June 13th to 19th, 2005.[3] Twenty-four teams competed in the four-legged league and were divided into eight groups of three for a round robin competition. The top 16 teams then moved on to a second round robin with 4 teams in each group to determine the top two teams which would advance to the quarterfinals. The teams in our initial group were JollyPochie from Kyushu University and Tohoku University in Japan; and UChile from Universidad de Chile. After finishing first in our group, we advanced to the second round robin in a group with CMDash from Carnegie Mellon University, EagleKnights from ITAM in Mexico, and BabyTigers from Osaka University. After finishing 2nd in that group, we advanced to the quarter-finals against rUNSWift from UNSW in Australia. The results of our six games are shown in Table E.7. Links to videos from these games are available at http://www.cs.utexas.edu/~AustinVilla/?p=competitions/roboCup_2005.

TABLE E.7: The Scores of Our Six Games at RoboCup

OPPONENT	SCORE (US–THEM)	NOTES
JollyPochie	3–0	
UChile	2–0	
CMDash	1–2	
EagleKnights	9–0	
BabyTigers	3–0	
rUNSWift	1–7	Quarterfinal

The most exciting game was our first official matchup against CMU. It was a very close game, with UT Austin Villa taking a 1–0 lead before eventually losing 2–1. Meanwhile, the match against rUNSWift demonstrated clearly that our team was not quite at the top level of the competition. Indeed two other teams (NuBots and the German Team) were also clearly stronger than UT Austin Villa.

[3]http://www.robocup2005.org.

References

[1] M. Ahmadi and P. Stone, "Continuous area sweeping: A task definition and initial approach," in *12th Int. Conf. Advanced Robotics*, July 2005.

[2] H. L. Akın, Ç. Meriçli, T. Meriçli, K. Kaplan and B. Çelik, "Cerberus'05 team report," Technical report, Artificial Intelligence Laboratory, Department of Computer Engineering, Boğaziçi University, Oct. 2005.

[3] M. Asada and H. Kitano, Eds. *RoboCup-98: Robot Soccer World Cup II*. Lecture Notes in Artificial Intelligence, vol. 1604. Berlin: Springer Verlag, 1999.

[4] J. A. Bagnell and J. Schneider, "Autonomous helicopter control using reinforcement learning policy search methods," in *Int. Conf. Robotics and Automation*, IEEE Press, 2001, pp. 1615–1620.

[5] S. Belongie, J. Malik and J. Puzicha, "Shape matching and object recognition using shape contexts," *Pattern Analysis and Machine Intelligence*, April 2002.

[6] A. Birk, S. Coradeschi and S. Tadokoro, Eds., *RoboCup-2001: Robot Soccer World Cup V*. Berlin: Springer Verlag 2002.

[7] G. Buchman, D. Cohen, P. Vernaza and D. D. Lee, "UPennalizers 2005 Team Report," Technical report, School of Engineering and Computer Science, University of Pennsylvania, 2005.

[8] S. Chen, M. Siu, T. Vogelgesang, T. F. Yik, B. Hengst, S. B. Pham and C. Sammut, *RoboCup-2001: The Fifth RoboCup Competitions and Conferences*. Berlin: Springer Verlag, 2002.

[9] ——, "The UNSW RoboCup 2001 Sony legged league team," Technical report, University of New South Wales, 2001. Available at http://www.cse.unsw.edu.au/~robocup/2002site/.

[10] W. Chen, "Odometry calibration and gait optimisation," Technical report, The University of New South Wales, School of Computer Science and Engineering, 2005.

[11] S. Chernova and M. Veloso, "An evolutionary approach to gait learning for four-legged robots," in *Proc. of IROS'04*, Sept. 2004.

[12] D. Cohen, Y. H. Ooi, P. Vernaza and D. D. Lee. *RoboCup-2003: The Seventh RoboCup Competitions and Conferences*. Berlin: Springer Verlag, 2004.

[13] ——, "The University of Pennsylvania RoboCup 2004 legged soccer team, 2004." Available at URL http://www.cis.upenn.edu/robocup/UPenn04.pdf.

[14] D. Comaniciu and P. Meer, "Mean shift: A robust approach toward feature space analysis," *IEEE Trans. Pattern Anal. Mach. Intell.*, vol. 24, No. 5, 2002, pp. 603–619.doi:10.1109/34.1000236

[15] R. O. Duda, P. E. Hart and D. G. Stork, *Pattern Classification*. John Wiley and Sons, Inc., 2001.

[16] U. Dueffert and J. Hoffmann, "Reliable and precise gait modeling for a quadruped robot," in *RoboCup 2005: Robot Soccer World Cup IX, Lecture Notes in Artificial Intelligence*. Springer, 2005.

[17] H. Work et al., "The Northern Bites 2007 4-Legged Robot Team," Technical report, Department of Computer Science, Bowdoin College, Feb. 2007.

[18] M. Quinlan et al., "The 2005 NUbots Team Report," Technical report, School of Electrical Engineering and Computer Science, The University of Newcastle, Nov. 2005.

[19] S. Thrun et al., "Probabilistic algorithms and the interactive museum tour-guide robot minerva," *Int. J. Robot. Res.*, vol. 19, No. 11, pp. 972–999, 2000. doi:10.1177/02783640022067922

[20] T. Rofer et al., "Germanteam 2005 team report," Technical report, Oct. 2005.

[21] W. Uther et al., "Cm-pack'01: Fast legged robot walking, robust localization, and team behaviors," in *Fifth Int. RoboCup Symp.*, 2001.

[22] F. Farshidi, S. Sirouspour and T. Kirubarajan, "Active multi-camera object recognition in presence of occlusion," in *IEEE Int. Conf. Intelligent Robots and Systems (IROS)*, 2005.

[23] P. Fidelman and P. Stone, "The chin pinch: A case study in skill learning on a legged robot," in G. Lakemeyer, E. Sklar, D. Sorenti and T. Takahashi, Eds., *RoboCup-2006: Robot Soccer World Cup X*. Berlin: Springer Verlag, 2007. Submitted for publication.

[24] G. Finlayson, S. Hordley and P. Hubel, "Color by correlation: A simple, unifying framework for color constancy," *IEEE Trans. Pattern Anal. Mach. Intell.*, vol. 23, No. 11, Nov. 2001. doi:10.1109/34.969113

[25] D. Forsyth, "A novel algorithm for color constancy," *Int. J. Comput. Vis.*, vol. 5, No. 1, pp. 5–36, 1990.doi:10.1007/BF00056770

[26] D. Fox, "Adapting the sample size in particle filters through kld-sampling," *Int. J. Robot. Res.*, 2003.

[27] D. Fox, W. Burgard, H. Kruppa and S. Thrun, "Markov localization for mobile robots in dynamic environments," *J. Artif. Intell.*, vol. 11, 1999.

[28] A. L. N. Fred and A. K. Jain, "Robust data clustering," in *Int. Conf. Computer Vision and Pattern Recognition*, pp. 128–136, June 2003.

[29] R. C. Gonzalez and R. E. Woods, *Digital Image Processing*. Prentice Hall, 2002.

[30] J.-S. Gutmann and D. Fox, "An experimental comparison of localization methods continued," in *IEEE Int. Conf. Intelligent Robots and Systems*, 2002.

[31] J.-S. Gutmann, T. Weigel and B. Nebel, "A fast, accurate and robust method for self localization in polygonal environments using laser range finders," *Advanced Robotics*, vol. 14, No. 8, 2001, pp. 651–668. doi:10.1163/156855301750078720

[32] B. Hengst, D. Ibbotson, S. B. Pham and C. Sammut, "Omnidirectional motion for quadruped robots," in A. Birk, S. Coradeschi and S. Tadokoro, Eds, *RoboCup-2001: Robot Soccer World Cup V*. Berlin: Springer Verlag, 2002.

[33] G. S. Hornby, M. Fujita, S. Takamura, T. Yamamoto and O. Hanagata, "Autonomous evolution of gaits with the Sony quadruped robot," in W. Banzhaf, J. Daida, A. E. Eiben, M. H. Garzon, V. Honavar, M. Jakiela and R. E. Smith, Eds, *Proc. Genetic and Evolutionary Computation Conf.*, vol. 2: Morgan Kaufmann: Orlando, Florida, USA, 1999, pp. 1297–1304.

[34] G. S. Hornby, S. Takamura, J. Yokono, O. Hanagata, T. Yamamoto and M. Fujita, "Evolving robust gaits with AIBO," in *IEEE Int. Conf. Robotics and Automation*, 2000, pp. 3040–3045.

[35] L. Iocchi, D. Mastrantuono and D. Nardi, "A probabilistic approach to hough localization," in *IEEE Int. Conf. Robotics and Automation*, 2001.

[36] A. K. Jain and R. C. Dubes, *Algorithms for Clustering Data*. Prentice Hall, 1988.

[37] P. Jensfelt, J. Folkesson, D. Kragic and H. I. Christensen, "Exploiting distinguishable image features in robotic mapping and localization," in *The European Robotics Symp. (EUROS)*, 2006.

[38] R. E. Kalman, "A new approach to linear filtering and prediction problems," *Trans. ASME, J. Basic Eng.*, vol. 82, pp. 35–45, Mar. 1960.

[39] G. A. Kaminka, P. U. Lima and R. Rojas, Eds., *RoboCup-2002: Robot Soccer World Cup VI*. Berlin: Springer Verlag, 2003.

[40] M. S. Kim and W. Uther, "Automatic gait optimisation for quadruped robots," in *Australasian Conf. Robotics and Automation*, Brisbane, Dec. 2003.

[41] H. Kitano, Ed. *RoboCup-97: Robot Soccer World Cup I*. Berlin: Springer Verlag, 1998.

[42] H. Kitano, M. Asada, Y. Kuniyoshi, I. Noda and E. Osawa, "RoboCup: The robot world cup initiative," in *Proc. First Int. Conf. Autonomous Agents*, Marina Del Rey: California, Feb. 1997, pp. 340–347. doi:10.1145/267658.267738

[43] N. Kohl and P. Stone, "Machine learning for fast quadrupedal locomotion," in *Nineteenth National Conf. on Artificial Intelligence*, July 2004, pp. 611–616.

[44] ——, "Policy gradient reinforcement learning for fast quadrupedal locomotion," in *Proc. IEEE Int. Conf. Robotics and Automation*, May 2004.

[45] C. Kwok and D. Fox, "Map-based multiple model tracking of a moving object," in *Int. RoboCup Symp.*, Lisbon, 2004.

[46] S. Lenser and M. Veloso, "Sensor resetting localization for poorly modelled mobile robots," in *Int. Conf. Robotics and Automation*, April 2000.

[47] J. Leonard and H. Durrant-Whyte, "Mobile robot localization by tracking geometric features," *IEEE Trans. on Robot. Autom.*, 1991.

[48] F. Lu and E. Milos, "Robust pose estimation in unknown environments using 2d range scans," *J. Intell. Robot. Syst.*, 18, 1997. doi:10.1023/A:1007957421070

[49] R. Madhavan, K. Fregene and L. E. Parker, "Distributed heterogenous outdoor multi-robot localization," in *Int. Conf. Robotics and Automation (ICRA)*, 2002.

[50] M. Montemerlo, S. Thrun, H. Dahlkamp, D. Stavens, and S. Strohband, "Winning the DARPA Grand Challenge with an AI robot, in *Proc. AAAI National Conf. Artificial Intelligence*, Boston, MA, July 2006.

[51] R. R. Murphy, J. Casper and M. Micire, "Potential tasks and research issues of mobile robots in robocup rescue, in P. Stone, T. Balch and G. Kraetszchmar, Eds., *RoboCup-2000: Robot Soccer World Cup IV* Berlin: Springer Verlag, 2001, pp. 339–344.

[52] D. Nardi, M. Riedmiller and C. Sammut, Eds., *RoboCup-2004: Robot Soccer World Cup VIII*. Berlin: Springer Verlag, 2005.

[53] A. Y. Ng, H. J. Kim, M. I. Jordan and S. Sastry, "Autonomous helicopter flight via reinforcement learning," in *Advances in Neural Information Processing Systems*, vol. 17 MIT Press Submitted for publication. To Appear.

[54] I. Noda, A. Jacoff, A. Bredenfeld and Y. Takahashi, Eds., *RoboCup-2005: Robot Soccer World Cup IX*. Berlin: Springer Verlag, 2006.

[55] C. Pantofaru and M. Hebert, "A comparison of image segmentation algorithms, cmu-ri-tr-05-40. Technical report, Robotics Institute, Carnegie Mellon University, September 2005.

[56] L. E. Parker, "Distributed algorithms for multi-robot observation of multiple moving targets," *Auton. Robots*, vol. 12, No. 3, pp. 231–255, 2002. doi:10.1023/A:1015256330750

[57] D. Polani, B. Browning, A. Bonarini and K. Yoshida, Eds., *RoboCup-2003: Robot Soccer World Cup VII*. Berlin: Springer Verlag, 2004.

[58] F. K. H. Quek, "An algorithm for the rapid computation of boundaries of run-length encoded regions," *Pattern Recognit. J.*, vol. 33, pp. 1637–1649, 2000. doi:10.1016/S0031-3203(98)00118-6

[59] M. J. Quinlan, S. K. Chalup and R. H. Middleton, "Techniques for improving vision and locomotion on the Sony AIBO robot," in *Proc. 2003 Australasian Conf. Robotics and Automation*, Dec. 2003.

[60] M. J. Quinlan, S. P. Nicklin, K. Hong, N. Henderson, S. R. Young, T. G. Moore, R. Fisher, P. Douangboupha and S. K. Chalup, "The 2005 nubots team report," Technical report, The University of Newcastle, School of Electrical Engineering and Computer Science, 2005.

[61] T. Roefer, R. Brunn, S. Czarnetzki, M. Dassler, M. Hebbel, M. Juengel, T. Kerkhof, W. Nistico, T. Oberlies, C. Rohde, M. Spranger and C. Zarges, "Germanteam 2005," in *RoboCup 2005: Robot Soccer World Cup IX, Lecture Notes in Artificial Intelligence.* Springer, 2005.

[62] T. Roefer, "Evolutionary gait-optimization using a fitness function based on proprioception"

[63] T. Rofer, H.-D. Burkhard, U. Duffert, J. Hoffman, D. Gohring, M. Jungel, M. Lotzach, O. v. Stryk, R. Brunn, M. Kallnik, M. Kunz, S. Petters, M. Risler, M. Stelzer, I. Dahm, M. Wachter, K. Engel, A. Osterhues, C. Schumann and J. Ziegler, "Germanteam robocup 2003," Technical report, 2003.

[64] T. Rofer and M. Jungel, "Vision-based fast and reactive Monte-Carlo Localization," in *IEEE Int. Conf. Robotics and Automation*, Taipei, Taiwan, 2003, pp. 856–861.

[65] C. Rosenberg, M. Hebert and S. Thrun, "Color constancy using kl-divergence," in *IEEE Int. Conf. Computer Vision*, 2001.

[66] C. Sammut, W. Uther and B. Hengst, "rUNSWift 2003 Team Report. Technical report, School of Computer Science and Engineering, University of New South Wales, 2003.

[67] Robert J. Schilling, *Fundamentals of Robotics: Analysis and Control.* Prentice Hall, 2000.

[68] A. Selinger and R. C. Nelson, "A perceptual grouping hierarchy for appearance-based 3d object recognition," *Comput. Vis. Image Underst.*, vol. 76, No. 1, pp. 83–92, Oct. 1999. doi:10.1006/cviu.1999.0788

[69] J. Shi and J. Malik, "Normalized cuts and image segmentation," in *IEEE Transactions on Pattern Analysis and Machine Intelligence (PAMI)*, 2000.

[70] M. Sridharan, G. Kuhlmann and P. Stone, "Practical vision-based Monte Carlo Localization on a legged robot," in *IEEE Int. Conf. Robotics and Automation*, April 2005.

[71] M. Sridharan and P. Stone, "Towards on-board color constancy on mobile robots," in *First Canadian Conf. Computer and Robot Vision*, May 2004.

[72] ——, "Autonomous color learning on a mobile robot," in *Proc. Twentieth National Conf. Artificial Intelligence*, July 2005.

[73] ——, "Real-time vision on a mobile robot platform," in *IEEE/RSJ Int. Conf. Intelligent Robots and Systems*, Aug. 2005.

[74] ——, "Towards illumination invariance in the legged league," in D. Nardi, M. Riedmiller and C. Sammut, Eds, *RoboCup-2004: Robot Soccer World Cup VIII, Lecture Notes in*

Artificial Intelligence, vol. 3276, Berlin: Springer Verlag, 2005, pp. 196–208.

[75] ——, "Towards eliminating manual color calibration at RoboCup," in I. Noda, A. Jacoff, A. Bredenfeld and Y. Takahashi, Eds., *RoboCup-2005: Robot Soccer World Cup IX*, vol. 4020. Berlin: Springer Verlag, 2006, pp. 673–381. doi:10.1007/11780519_68

[76] ——, "Autonomous planned color learning on a legged robot," in G. Lakemeyer, E. Sklar, D. Sorenti and T. Takahashi, Eds., *RoboCup-2006: Robot Soccer World Cup X*. Berlin: Springer Verlag, 2007. Submitted for publication.

[77] P. Stone, T. Balch and G. Kraetzschmar, Eds., *RoboCup-2000: Robot Soccer World Cup IV, Lecture Notes in Artificial Intelligence*, vol. 2019, Berlin: Springer Verlag, 2001.

[78] P. Stone, K. Dresner, S. T. Erdoğan, P. Fidelman, N. K. Jong, N. Kohl, G. Kuhlmann, E. Lin, M. Sridharan, D. Stronger and G. Hariharan, "UT Austin Villa 2003: A new RoboCup four-legged team," Technical Report UT-AI-TR-03-304, The University of Texas at Austin, Department of Computer Sciences, AI Laboratory, 2003.

[79] P. Stone, K. Dresner, P. Fidelman, N. K. Jong, N. Kohl, G. Kuhlmann, M. Sridharan and D. Stronger, "The UT Austin Villa 2004 RoboCup four-legged team: Coming of age," Technical Report UT-AI-TR-04-313, The University of Texas at Austin, Department of Computer Sciences, AI Laboratory, Oct. 2004.

[80] P. Stone, K. Dresner, P. Fidelman, N. Kohl, G. Kuhlmann, M. Sridharan and D. Stronger, "The UT Austin Villa 2005 RoboCup four-legged team," Technical Report UT-AI-TR-05-325, The University of Texas at Austin, Department of Computer Sciences, AI Laboratory, Nov. 2005.

[81] P. Stone, P. Fidelman, N. Kohl, G. Kuhlmann, T. Mericli, M. Sridharan and S. E. Yu, "The UT Austin Villa 2006 RoboCup four-legged team," Technical Report UT-AI-TR-06-337, The University of Texas at Austin, Department of Computer Sciences, AI Laboratory, Dec. 2006.

[82] D. Stronger and P. Stone, "A model-based approach to robot joint control," in D. Nardi, M. Riedmiller and C. Sammut, Eds., *RoboCup-2004: Robot Soccer World Cup VIII, Lecture Notes in Artificial Intelligence*, vol. 3276, Berlin: Springer Verlag, 2005, pp. 297–306.

[83] ——, "Simultaneous calibration of action and sensor models on a mobile robot," in *IEEE Int. Conf. Robotics and Automation*, April 2005.

[84] A. W. Stroupe, M. C. Martin and T. Balch, "Merging probabilistic observations for mobile distributed sensing," Technical Report CMU-RI-00-30, Carnegie Mellon University, Pittsburgh, PA, 2000.

[85] B. Sumengen, B. S. Manjunath and C. Kenney, "Image segmentation using multi-region stability and edge strength," in *IEEE Int. Conf. Image Processing (ICIP)*, Sept. 2003.

[86] S. Thrun, "Particle filters in robotics," in *17th Annual Conf. on Uncertainty in AI (UAI)*, 2002.

[87] S. Thrun, D. Fox, W. Burgard and F. Dellaert, "Robust Monte Carlo Localization for mobile robots," *J. Artif. Intell.*, 2001.

[88] A. Torralba, K. P. Murphy and W. T. Freeman, "Sharing visual features for multiclass and multiview object detection," in *IEEE Conf. Computer Vision and Pattern Recognition (CVPR)*, Washington DC, 2004.

[89] M. Veloso, S. Lenser, D. Vail, M. Roth, A. Stroupe and S. Chernova, "CMPack-02: CMU's legged robot soccer team," Oct. 2002.

[90] M. Veloso, E. Pagello and H. Kitano, Eds., *RoboCup-99: Robot Soccer World Cup III*. Berlin: Springer Verlag, 2000.

[91] R. Zhang and P. Vadakkepat, "An evolutionary algorithm for trajectory based gait generation of biped robot, in *Proc. Int. Conf. Computational Intelligence, Robotics and Autonomous Systems*, Singapore, 2003.

Biography

Dr. Peter Stone is an Alfred P. Sloan Research Fellow and Assistant Professor in the Department of Computer Sciences at the University of Texas at Austin. He received his Ph.D. in 1998 and his M.S. in 1995 from Carnegie Mellon University, both in Computer Science. He received his B.S. in Mathematics from the University of Chicago in 1993. From 1999 to 2002 he was a Senior Technical Staff Member in the Artificial Intelligence Principles Research Department at AT&T Labs - Research.

Prof. Stone's research interests include planning, machine learning, multiagent systems, robotics, and e-commerce. Application domains include robot soccer, autonomous bidding agents, traffic management, and autonomic computing. His doctoral thesis research contributed a flexible multiagent team structure and multiagent machine learning techniques for teams operating in real-time noisy environments in the presence of both teammates and adversaries. He has developed teams of robot soccer agents that have won six robot soccer tournaments (RoboCup) in both simulation and with real robots. He has also developed agents that have won four auction trading agent competitions (TAC). Prof. Stone is the author of "Layered Learning in Multiagent Systems: A Winning Approach to Robotic Soccer" (MIT Press, 2000). In 2003, he won a CAREER award from the National Science Foundation for his research on learning agents in dynamic, collaborative, and adversarial multiagent environments. In 2004, he was named an ONR Young Investigator for his research on machine learning on physical robots. Most recently, he was awarded the prestigious IJCAI 2007 Computers and Thought award.

Printed in the United States
by Baker & Taylor Publisher Services